Human Perspectives in Health Sciences and Technology

Volume 150

The Human Perspectives in Health Sciences and Technology series publishes volumes that delve into the coevolution between technology, life sciences, and health sciences. The distinctive mark of the series is a focus on the human, as a subject and object of research. The series provides an editorial forum to present both scientists' cutting-edge proposals in health sciences that are able to deeply impact our human biological, emotional and social lives and environments, and thought-provoking theoretical reflections by philosophers and scientists alike on how those scientific achievements affect not only our lives, but also the way we understand and conceptualize how we produce knowledge and advance science, so contributing to refine the image of ourselves as human knowing subjects and active participants in a constantly evolving environment. The series addresses ethical issues in a unique way, i.e. an ethics seen not as an external limitation on science, but as internal to scientific practice itself; as well as an ethics characterized by a positive attitude towards science, trusting the history of science and the resources that, in science, may be promoted in order to orient science itself towards the common good for the future. This is a unique series suitable for an interdisciplinary audience, ranging from philosophers to ethicists, from bio-technologists to epidemiologists as well to public health policy makers.

Adriano Fabris • Sergio Belardinelli
Editors

Digital Environments and Human Relations

Ethical Perspectives on AI Issues

Springer

Editors
Adriano Fabris
Department of Civilization
and Forms of Knowledge
University of Pisa
Pisa, Italy

Sergio Belardinelli
Department of Political and Social Sciences
University of Bologna
Bologna, Italy

ISSN 2661-8915 ISSN 2661-8923 (electronic)
Human Perspectives in Health Sciences and Technology
ISBN 978-3-031-76960-3 ISBN 978-3-031-76961-0 (eBook)
https://doi.org/10.1007/978-3-031-76961-0

This Springer imprint is published by the registered company Springer Nature Switzerland AG
The registered company address is: Gewerbestrasse 11, 6330 Cham, Switzerland

Acknowledgements

We want to thank the Campus Bio-Medico University of Roma in the persons of its last two rectors, Raffaele Calabrò and Eugenio Guglielmelli, for promoting an internal reflection within the University on issues pertaining to the development and future of the human being. They have also done so by establishing and supporting the activities of an interdisciplinary Consulting Committee. This Committee includes—in addition to the two editors of this volume, Sergio Belardinelli as chairman and Adriano Fabris as a member—Felice Barela, Andrea Maccarini, Giulio Maspero, Pablo Requena and Paola Ricci Sindoni. This book is also the result of a series of activities—a seminar and an international conference—conceived and promoted by the Committee. Thanks then to Chiara Carrozza, president of the CNR (National Research Council of Italy), who competently interacted with our project. And finally, thanks to Elio Grande who collaborated intelligently in editing the book.

Adriano Fabris
Sergio Belardinelli

Contents

Ethics and Digital Environments: Introduction

Adriano Fabris and Sergio Belardinelli

1 Physical Environments and Digital Environments

Nowadays, one of the hallmarks of our daily experience is the fact we are living in multiple environments. They are physical environments and digital environments. For the individual, the physical environment is just one, the one in which he moves with his own body, while digital environments can be more than one and depend on what telematic connections can offer (e.g. which platforms are used).

However, the dimension of multiplicity is not the only one that should be kept in mind when we speak of the environments we live in today. Actually, we can live in many of such environments all at the same time. Indeed. The very fact the notion of "environment" has been extended to the digital dimension is what enables all of us to make the most of the many possibilities they offer, generally all together. This is another one of the changes we have to reckon with and needs to be questioned, instead of being merely described.

Why do we speak of "environment" in all these instances? Why, in addition to the environment of "nature", of what is born and grows according to its own spontaneous reasonings, do we allude to digital contexts too with the very same word—"environment"? The answer is easy: because in all these instances such dimensions can be inhabited by humans. Or even better: because, in such dimensions, humans can feel at home and can themselves build more niches, more "homes" within them (Valera & Castilla, 2019).

A. Fabris (✉)
Department of Civilisation and Forms of Knowledge, University of Pisa, Pisa, Italy
e-mail: adriano.fabris@unipi.it

S. Belardinelli
Department of Political and Social Sciences, University of Bologna, Bologna, Italy
e-mail: sergio.belardinelli@unibo.it

A. Fabris, S. Belardinelli (eds.), *Digital Environments and Human Relations*,
Human Perspectives in Health Sciences and Technology 150,
https://doi.org/10.1007/978-3-031-76961-0_1

The idea of home helps us understand why the notion of "environment" has been extended from the natural dimension to the artificial one (Strate, 2017). Nature, for humans, can be felt as home only if it is cultivated, transformed, built: in other words, only if it turns into culture. Actually, the lived-in environment is always an artificial environment. Now, though, it is precisely such artificiality that has been specifically developed and extended. The artificial is no longer the mere result of a technical work, it is above all the result of a technological change. It is technology that provides humans with more and more environments to live in. They can because they have an increasing level of autonomy and independence from human action, even when such action is enhanced by the use of technical devices (Fabris, 2018).

Let's look, for instance, at a specific category of technological environments: the one related to information and communication technology (ICT). Communication today is something that is not only done, it is mainly experienced. It is not purely an activity that is performed or suffered by humans, and to a certain specific extent by some artificial entities too, but one that helps build the world we move in and interact with.

Today, "communicating" no longer means just transmitting data from a transmitter to a receiver (based on the concept worked out by Shannon & Weaver, 1949) or finding a way to live more comfortably in a world that is a given through the relations (with the world and with other humans) that all the different types of communication can build. Today, the devices that transmit such data—such as computers, smartphones, or networks of devices –, as well as the programmes through which this happens, help create specific worlds and give access to further relational contexts, beyond the physical and cultural ones we normally live in. They do that just by transmitting data.

This is a crucial point. It is precisely data transmission, in its pervasiveness and fastness, that ends up having an additional effect on reaching a recipient more or less effectively and efficiently. Something else comes to the fore, here. Data transmission involves creating environments in which we can move quite similarly to the way we do in the contexts we physically inhabit with our bodies.

In these cases, what the transmission of a message conveys is the opening, maintenance, development of a set of relations. It is what is expressed by the word "infosphere", which is so popular nowadays. Infosphere is the set of informational entities in which we are immersed and their mutual interactions. More precisely, it is that environment, organised by data and information, in which we live and with which we constantly interact, both offline and online, similarly to what the biosphere is, as the realm in which living beings operate (Floridi, 2016).

There's more than this. Infosphere—the communicative environment we move in—is not made up of data and information only, of their connections and their constant, growing and superfast ability to transmit them. Just because it is the environment we live in, it stands out for something else. It consists of structures of meaning that can involve us, and not just of information or procedures to transmit it. And such structures are built and implemented by all the entities that are involved in the communicative activity: be they natural or artificial, humans or machines operating within them.

Only in connection with such meaningfulness, if you think about it, can we speak of an "environment" proper, that is, a "home" in which we can find our orientation (Kant, 1991) and which we can properly inhabit. A home is made of meaningful objects and, as such, it becomes meaningful to us. So, only a communicative environment can attract us more than some other one. Only on account of its meaningfulness can we be driven to move on from a—physical or digital—context to a different one.

After all, this does not mean that such meaningfulness is created by us, as users, or that, conversely, it is imposed on us within a given context, whether natural or artificial. It is instead the effect of our expectations—the expectations of a being, the human being, who always outdoes the world—it is the convergence of what the world offers from time to time, what the world shows us that is more or less interesting and engaging for us at a given time, and our ability to grasp it. And this applies to digital environments as well. It is us who interpret and give a social value to some given programmes, using them as relational places. It is us who recognise such value, who call on and implement it. But all this is possible because a given programme is built in a given way and therefore it can itself meet certain needs or trigger certain instincts we may have. We can check that out, for instance, in the structure of Social Networks.

2 An Ethical Perspective

In this scenario, just describing the processes that involve us is not enough. We should ask ourselves, instead, how we can deal with such processes. We need to examine closely some of the ethical implications of what we have just said. This is about the way we can govern our habitats in these new environments. More precisely, this is about the ways in which we can act well and feel at home in each one of them, and moving from one to the other.

The key question is first and foremost making a distinction between the different environments we inhabit and relating them with each other, visiting any of them consciously and competently, living in them in a good way. Then, the question more precisely concerns, on one hand, the need to properly deal with such environments, in all their variety, and on the other hand to move ethically within them. This is a topic that in fact certainly concerns every one of our ways of experiencing an environment, be it physical or digital, but one that also takes on a special relevance and urgency in the current technological contexts.

One can actually experience the coexistence of such environments, the physical and the digital ones, and one can move between them. In a physical context, one needs to find one's bearings, first. The same happens in a digital dimension. Then, if many environments are possible, one needs to choose the favourite ones and arrange them in a hierarchy. If they are both physical and digital environments, such orientation and such choice cover different types of experiences.

If the natural, cultural and technological online and offline environments are many, whether they are synergistically connected with each other or mutually conflicting, one needs not only assess each one of them and find the best way to visit them, but also learn to navigate this new, complex scenario. In other words, one needs to hierarchise the spaces one can live in, rank them, and choose on a case-by-case basis the most suitable one for the activity one is performing. The perspective of *an inter-environmental ethics* comes to the fore, here: between artificial environments on one side and physical and cultural environments, on the other side. From such perspective, then, ethics can help humans become—as it were—selectors and organisers of worlds.

The developments of AI have further complicated the question we are discussing. They have, because now those very same environments—the digital environments—can reproduce and implement themselves. This is the first point that must be kept in mind: the fact that, through some Artificial Intelligence programmes, some environments may autonomously develop and multiply, outside any specific human control. The *multiverse*—today, we talk more specifically of "metaverse"—is by now a reality close at hand (Ball, 2024). In addition, the inhabitants of such environments, the entities acting within them—can even be—and are actually increasingly so—no longer human but artificial entities. They act more and more autonomously: just think of what the many chatbots do. We are increasingly called to interact with them. This is why an ethical reflection on this field is becoming increasingly urgent too.

3 The Chapters in This Volume

The works contained in this volume intend to explore many of the problems that such complex scenario entails. They have been written by authors from different backgrounds, who therefore offer, and have mutually compared, their reflections from an interdisciplinary perspective. They include philosophers, computer scientists, jurists, sociologists, scientists. Through their contribution, this volume intends to discuss matters that revolve around the ethics of technology (with a focus on AI ethics) from a distinctive perspective, by bringing the reflection back to the factual realm of our life in digital environments and raising the question of humans' orientation and wellbeing within such environments. The perspective shared by all these papers is, in a nutshell, a self-confessedly humanistic perspective. The human being is and remains at the centre, just because the human being is the only one who can ask himself ethical questions when faced with such situations and such actions.

The paper of Adriano Fabris, Silvia Dadà and Elio Grande (*Towards a Relational Ethics in AI*) intends to discuss some of the core problems of AI ethics. If we assume the AI ethics to be an ethics of relations between human and artificial agents, then first we must explain what "acting" means in both cases, then outline the different forms of relationships that can be built and regulated between human and artificial agents, in ethical terms and with a deontological approach. For this purpose, an insight into the codes of law and documents produced across the world, including their scope of applicability and global acceptance, becomes essential. Equally important is laying down the right ways to build such relations, not only through a procedure but by designing a system that can develop and implement a common, cooperative and responsible conduct for humans and artificial entities alike. The three paragraphs which this Chapter is divided into will provide an insight into those three sets of problems.

The Chapter of Giampaolo Ghilardi and Silvia Iadaresta (*Ethics of Digital Technologies: Where we are, Where we are heading towards*) aims to explore the anthropological and moral implications of the new landscape digital technologies are bringing about. To achieve this task, the paper will explore the change that the notion (and the reality) of environment has undergone, from a naturalistic concept to the actual synthetic one. Such analysis will lead to consider this new phenomenon that will be called "the primacy of potency on actuality", the opposite of the Aristotelian philosophy. The reasoning will step forward then to the idea of dataism. Dataism will be studied as the new ideology, or framework if the previous word sounds too strong, behind "digital life". Once it has assessed the nature of dataism, the paper will examine another important tool for this new environment: the macroscope, that is, the theoretical instrument enabling the user to capture layers of reality through Data correlation, otherwise not available. After discussing another two main concepts—the concepts of "algorethics" and of "on-life" –, the Chapter goes on to explore the real ethical issue that needs to be addressed: how it is possible, in such environment, for the human being to thrive and even rejoice, in a word: to flourish? To answer this question, the text finally develops a new perspective on "human ecology".

The following essay, by Giacinto Barresi (*Neuroergonomics for Human-Technology Interaction Design and Assessment*), focuses on the discipline of ergonomics and, more specifically, on neuro-ergonomics. Ergonomics investigates how a human being interacts with other elements of a system (in this case, of a technological context). Simultaneously, this discipline also aims at improving the wellbeing of the person and the performance of the system itself. Each branch of ergonomics (physical, cognitive, organizational) offers valuable methodologies for studying the space where human-technology interactions happen and show their effects. In particular, as we said, in his text Barresi will focus on the contributions of neuroergonomics, the discipline investigating human factors according to neuroscientific methods. Such an approach can exploit the rich data collected, for instance, by wearable devices in ecologically valid contexts based on physical and digital (and phygital) spaces populated by humans and smart machines, making the design of the latter truly person-centred.

The fifth work (Carlo Casonato, Giulia Olivato, *The Challenges of AI Act*) is split into two parts. The first part summarises the characteristics that AI law should have to regulate a sensitive, pervasive, and transformative area such as AI in an effective and fair way; and considers whether the text of the AI Act meets these requirements. The second part examines the main contents of the AI Act itself. Both the positive features and the complexities and difficulties of such documents, in terms of enforcement, are highlighted.

Stefano Cozzini and Mariarita De Luca's Chapter (*Living in Digital Ecosystems. Are we aware of this?*) discusses and assesses the social and economic impact of contemporary digital ecosystems. The authors extend the definition of digital ecosystem, inspired by the idea of circular economy, based on an enhanced and efficient use/reuse of resources and products. They consider an Extended or Circular Digital Ecosystem, a Digital Ecosystem in which the final customers can create value by sharing their data and digital contents through the digital ecosystem services themselves. In this way, data consumers also become data producers. This new concept of a digital ecosystem is closely integrated with the role and actions of the people involved in it, as users, producers, and managers at the same time. In this context, the Chapter addresses the power of AI algorithms and shows that such problem does not only concern algorithms *per se* but also concerns the data used to train them. The users of digital ecosystems should be aware that all available data are currently used to train AI algorithms and the results that these algorithms produce depend on the quality of the data used to train them. It has been widely demonstrated that data classification by human intervention is prone to biases and errors. Algorithms trained on those datasets will suffer from these biases, resulting in dangerous or inappropriate content. From this point of view, it is crucial to consider the technical experience of data managers and curators in the wider context of data-sharing and -management in society, given the immense amount of data produced and available on line. Users of digital ecosystems should also be made aware that a proper sharing of their information can improve the quality of the data available on line and that the potential dangers of data misuse can be reduced.

The Chapter written by Pecchia, Maccaro, Matarrese, Folkvord and Fico (*Navigating Digital Environments: Proximal Ignorance and Science's Responsibility*) proposes a parallel between important EU legal frameworks, such as the Medical Device Regulations and General Data Protection Regulation, and economic-historical challenges faced by European citizens. This parallel offers a perspective for understanding the direction taken by policymakers for forthcoming regulations, such as the European Health Data Space and the AI-Act. The paper also describes key commonalities and differences among those regulations, which may help scholars working with AI in medicine navigate such a far-ranging legal framework. it should be ensured that fair data-sharing and a proper use of AI are crucial to sustainable development, peace, and democracy in Europe and beyond.

In their Chapter (*Prudential Governance of AI*), Vittoradolfo Tambone and Gianmarco Di Palma address precisely the possibilities and the ways to govern technological advancements. In particular, they intend to shed light on and discuss the meta-project that underlies the current developments of AI and that can give consistency to its goals. After clarifying such point, the authors delve into an analysis of the virtue of prudence, meant as a virtue to call on to govern the consequences of the developments of AI on a social political and economic scale. This approach to such developments mainly takes inspiration from some key ethical principles: the intentional structure of human action, i.e. the need to connect to an absolute meaning, the interweaving of a project with a science at the service of mankind, the association with the normative force of natural law. The Chapter is closed by an analysis of the guiding principles of the United Nations—Interim Report: *Governing AI for Humanity*, and the 17 Sustainable Development Goals proposed by the United Nations to better connect a prudential AI governance with the common good.

The Chapter written by Roger Strand (*Beyond Digital Escapism*) brings us to further expand our horizon on what has been investigated so far. Strand starts by analysing two key types of discourse that are now the preferential ones: the one about "Geos" and the Anthropocene, with a look to the technological advancements in human history and their devastating technological impact, and the one about "Bios" and the transformation of our lives by the increasingly massive use of digital technology. Strand's idea, which also involves a task that intellectuals cannot shy away from, is reviving the open attitude that is typical of hope. As to the specific issue of digital technology, the question is not so much about whether it may be described as part of the human environment but whether and how it may be used to serve those beings that live in the real world.

The Volume is closed by Alfredo Marcos and Marta Bertolaso's Chapter (*Technological Humanism & Sustainable AI*). In their paper, Marcos and Bertolaso argue that what actually exists is not AI as such but *AI systems* that unfold from the combination and dynamic interactions of human beings and machines. In this sense, digitally-mediated environments deserve attention, philosophical and scientific analysis as well. In other words, it is necessary to move on, from considering AI systems as technical systems with social consequences to considering them as technically-implemented social systems. People are part of AI systems, as designers and users. As a consequence, in AI systems the intelligent part is not artificial, and the artificial part is not intelligent. The question about sustainable AI is thus not a technical problem, but an ontological and anthropological one, a problem of human ecology. In conclusion, from an ethical viewpoint, the nature and the virtues that human beings can develop in environments that are also digitally mediated opens up new scenarios for education, policies and innovation that has to be sustainable by design.

All things considered, these Chapters disclose elements that are worth reflecting upon. All of them make a factual contribution to the current debate on the ethics of technological innovations, especially those that revolve around AI. The interdisciplinary approach proves once again that a humanistic perspective is needed to deal with such problems. This means being aware of the fact that digital environments—that digital environments *too*—are inhabited by humans and have been made for humans to live in. This certainly entails new questions that need to be dealt with, and dealt with collaboratively. But it also requires a new awareness of what such changes involve, it requires moderation in addressing them, the willingness not to fall into line, with mere subservience in mind, and—first and foremost, as Strand points out—the development of an open and hopeful attitude towards the future. Only then can this milestone in human life be properly dealt with and managed, in pursuit of the common good.

References

Ball, M. (2024). *The Metaverse*. Liveright.

Fabris, A. (2018). *Ethics of ICTs*. Springer.

Floridi, L. (2016). *The fourth revolution. How the Infosphere is reshaping human reality*. Oxford University Press.

Kant, I. (1991). What is orientation in thinking? In H. S. Riess (Ed.), *Political writings* (pp. 235–249). Cambridge University Press.

Shannon, C., & Weaver, W. (1949). *The mathematical theory of communication*. The University of Illinois Press.

Strate, L. (2017). *Media ecology. An approach to understanding the human condition*. Lang.

Valera, L., & Castilla, J. C. (Eds.). (2019). *Global changes. Ethics, politics and environment in the contemporary technological world*. Springer.

Towards a Relational Ethics in AI. The Problem of Agency, The Search for Common Principles, the Pairing of Human and Artificial Agents

Adriano Fabris, Silvia Dadà, and Elio Grande

Abstract In this paper, some of the core problems of AI ethics are discussed. Firstly, assuming that AI ethics is to be an ethics of relations between human and artificial agents, we try to explain what "acting" means with respect to humans and AI systems. To do so, we outline the different forms of relationships that can be built and regulated between human and artificial agents, also according to a deontological approach. Secondly, then, we give an insight into the regulatory codes and documents produced across the world. We discuss their scope of applicability and global acceptance. In particular, we note that, along with principles of transparency, responsibility, privacy etc., some *relational* ethical principle is missing, although seeming essential. Thirdly, as a consequence, trying to concretely build upon such relational principle, we lay down the concept of a new kind of artificial intelligence systems design called *endless tuning*, which can help to develop and procedurally implement a common, cooperative and responsible conduct for humans and artificial entities alike.

Keywords AI ethics · Acting · Relational AI · AI principles · AI ethical codes · Cooperative design

A. Fabris (✉) · S. Dadà
Department of Civilisation and Forms of Knowledge, University of Pisa, Pisa, Italy
e-mail: adriano.fabris@unipi.it; silvia.dada@unipi.it

E. Grande
Department of Computer Science, University of Pisa, Pisa, Italy
e-mail: elio.grande@phd.unipi.it

A. Fabris, S. Belardinelli (eds.), *Digital Environments and Human Relations*, Human Perspectives in Health Sciences and Technology 150,
https://doi.org/10.1007/978-3-031-76961-0_2

1 Foreword

This paper intends to discuss some of the core problems of AI ethics. If we assume the AI ethics to be an ethics of relations between human and artificial agents, then first we must explain what "acting" means in both cases, then outline the different forms of relationships that can be built and regulated between human and artificial agents, in ethical terms and with a deontological approach. For this purpose, an insight into the regulatory codes and documents produced across the world, including their scope of applicability and global acceptance, becomes essential. Equally important is laying down the right ways to build such relations, not only through a procedure but by designing a system that can develop and implement a common, cooperative and responsible conduct for humans and artificial entities alike. The three paragraphs which this chapter is divided into will provide an insight into those three sets of problems.[1]

2 A Few Core-Issues

To properly address the problems of AI ethics, a few core issues need to be investigated from an epistemological perspective, first. They concern the way such a long-established discipline as ethics can properly relate to the new technological advances, such as those involved in what, with a quite apt metaphor, is called "Artificial Intelligence". Such questions are:

2.1 The need to be aware of what "acting" means for humans and artificial entities;
2.2. The need to take the right approach and the right distance when interacting with the so-called "Artificial Intelligence", avoiding easy anthropomorphisations and the risks inherent in the use of a metaphorical language;
2.3. The need to outline a general framework within which an AI ethics may be discussed and explored in all respects.

It is essential to engage in this preliminary reflection as, if such issues are not properly dealt with, there is the risk of discussing the phenomenon of AI ethics only superficially, confusingly and improperly. So, let's do it now, though briefly.

[1] This Chapter is part of the FAIR project (*Future of Artificial Intelligence Research*), in particular the *Artificial Intelligence: Foundational Aspects* research (FAIR—PE1; project no. PE00000013). The first of the following paragraphs was written by Adriano Fabris (§2), the second one by Silvia Dadà (§3), and the third one by Elio Grande (§ 4). The Conclusions (§ 5) were co-authored. The paragraph written by Elio Grande expands, revises and constitutes an in-depth analysis of the subject matter of a paper by the same author (no DOI accessible yet): Grande, E. Noi e lei. Si potrebbe entrare in sintonia con l'intelligenza artificiale? In Temperino, E., Tortoli, C. (Eds.). (2024). *Umani Artificiali. Studi e prospettive per una nuova quotidianità*. Pisa: ETS.

2.1 What Acting Means

Ethics deals with acting. It outlines criteria and principles for acting. Now, though, artificial devices act too. They act similarly to humans. They are entities that can quite reasonably be called "artificial moral beings" (Moor, 2006; Wallach & Allen, 2009; Misselhorn, 2022). However, in these particular cases, we need to explain what "acting" means even more than trying to understand in what sense we can speak of "moral acting" when speaking of artificial agents.

Actually, despite their similar results, there is no identity between human acting and the acting of AI programmes and devices. This is a point that must be very accurately highlighted, rejecting Turing's oversimplistic approach and giving priority to an ethical, not an ontological, approach to the matter. It must be done, because the chance to build a proper relationship between the human agent and the artificial agent, and therefore develop an AI ethics, depends on the acknowledgement of the peculiarities of the parties involved. As a matter of fact, acting means, first and foremost, interacting. And a relationship can happen only if there is a difference, not an identification, between the parties involved.

In the literature, the difference between natural agents and artificial agents is established either by the different degrees of independence of the agent's action or by the different types of agents. The latter are classified on the basis of whether its actions are goal-directed or intentional. In either case, we speak of a rationality of action, we lay down the conditions for it to be defined as "moral", and we try to find, not always successfully, the same distinctive features of moral rationality both in the human being and in the artificial entity (Misselhorn, 2022 [36 ss]).

I do not feel this can be a very fruitful route: both because many are the definitions of "intentionality" and because such quest, more or less explicitly, aims at finding the common features rather than the differences between the two types of agents. To develop a proper relationship, we need to get a better understanding of the differences, instead. To investigate them, then, I propose to add a category that features largely in the history of thought. We can regard human action a sort of "voluntary" action.

In common parlance, "voluntary" is all that stems from a wilful act. In ancient times, in the Greek world, "voluntary" meant something that was done spontaneously, readily, something that was second nature to someone. In the Judeo-Christian tradition, instead, such term stands for what is done intentionally, level-headedly, even in contrast with one's wishes, drives and natural urges. Lastly, first in the modern age and then in the contemporary mindset, willingness becomes absolute, rejects all constraints, believes it can give meaning to its own actions, affirms itself as such. In the first instance, then—if we want to add another term associated with the same conceptual dimension—one is free because one harmoniously conforms to one actually is; in the second instance, freedom is exercised by meeting those criteria and principles that, even regardless of our inclinations and "nature", steer us towards goodness; in the third instance, it is experienced in the rejection of any impositions from the outside and in unrestrained self-fulfilment.

The way in which we can conceptualise the action of machines is different. This most strikingly applies to Artificial Intelligence programmes or devices. As we saw before, these programmes and devices have their specific independence in the way they operate, insofar as they are programmed to interact with their environment. This means that they are technological, not merely technical, devices, since they can implement their actions in the world, in unpredicted and unpredictable ways, through a data-processing procedure that in many cases seems to be far from transparent. The disappearance of human control, and therefore the unpredictability I am speaking of, concerns the way some programmes and devices perform their operations, as well as the results and consequences thereof. This is what their distinctive "autonomy"—or, more precisely, "independence"—really means: the one that takes place in the machine-learning process.

This means that, as opposed to human actions, their activity may be considered "free" insofar as, though following specific algorithmic procedures, it can happen in unpredicted and unpredictable ways. This is "freedom from" human control, both in the way such procedures are performed and in their consequences. "Freedom to" do something is instead only partly present here, viz., it is insofar as the result that a programme can achieve can be attained because such programme has been designed by someone else, not by itself, for such purpose. As to "freedom for", which is a deliberate attempt to achieve specific consequences based on an equally deliberate assumption of specific principles that drive one to achieve them, it has no place in the AI dimension.

If "awareness" is taken to mean an agent's ability to explicitly assume the criteria and principles that guide its actions, and to retroact on them by changing its relationship with them or going so far as to change them, then, as things stand, we cannot attribute such property to programmes and devices. If anything, we can treat them *as if* they had it. But this is an example of that human projection I will address later. Actually, an artificial agent can only follow a procedure, unable to change its design but able to develop it.

This is the feature that distinguishes it from a human agent. More precisely, such difference lies *in the way* a given goal is achieved, not in *the fact* such goals can be achieved. In the second instance, as shown by the Turing test (Turing, 1950), there is a possibility of confusing human action with the action of a programme. But what is overlooked here is precisely the way in which an action is performed: voluntary (as we called it, according to tradition) and procedural, respectively. if we consider voluntary actions, which are initiated by the agent and can be re-determined by it, considerations and judgements are possible that are not required for procedural actions, since the only applicable criterion, decided since the beginning, is the effectiveness of the process.

With AI programmes and devices, let me say that again, we cannot speak of a "voluntary" act in the sense of an act that is controlled by the agent itself (in the sense of early, deliberate control over the consequences or in the sense of self-control over the decision whether to do something or not). Yet, this does not mean that the category of "voluntariness" will disappear from the conceptual vocabulary of our times. Two transformations come to the fore about it, instead, which make it applicable to artificial agents too: at the risk of potential mix-ups.

On one hand, the word "voluntary" means, of course, "spontaneous", yet "according to procedure": even when such procedure is implemented in a non-predetermined way or at least in a way that is unintelligible to humans. In other words, here, "spontaneity" happens to mean "according to one's structure", "according to the way a programme has been structured", even when it involves random consequences. On the other hand, instead, "voluntary" stands for a powerful, unbound will. Only that such will, meant at first as an enhancement and an extension of human action, soon turns out to be the distinctive feature of an autonomous entity. This bonds a given idea of power—conceived of according to the unit of measurement of a calculation and the parameters of efficiency—with a distinctive notion of initiative and autonomy, which humans, by embracing such idea, may even be prepared to bow to.

All this sounds misleading. It is true that, even after all the transformations of the technological language and advances, the action of artificial entities and human action risk being mixed up with each other. But, as we have seen, there are quite specific distinctive features—for instance, specific and different uses of the terms "intention", "voluntariness", freedom", or a focus on the ways rather than the goals of an action—which come to the fore in these circumstances. Only through such distinction can an AI ethics be worked out. But, before we can do that, we need to explain why such confusion may arise.

2.2 *The Anthropomorphic Projection*

This is an attitude that can hardly be taken, just because of a peculiar human aptitude and use of language that is specific to us, as beings provided with that ability to express ourselves. This aptitude tends to anthropomorphise, viz., to make entities look like ourselves though they are different, even radically different, from what we are, in order to understand them. The language through which we express such inclination reinforces and successfully accomplishes the identification (Fabris, 2016a). The result is a process of deceit, or better, of self-deceit (Natale, 2021).

We can see all this demonstrated by the now common use of the phrase "artificial intelligence". It is a metaphor, a label, which, just by expressing a general identification with human intelligence—despite the addition of the specific difference that refers to something "artificial" as obviously opposed to some ill-defined "natural" –, attracts public interest and also creates a polarisation between those who fear and those who, without a second thought, confide in devices endowed with such form of "intelligence". This results in a phenomenon of optic distortion due to the use of a metaphorical, inaccurate yet extremely evocative language.

An example of this now widespread approach can be found in one of the many definitions of AI that are currently available. This one was worked out by the EU to support the efforts made by a High Level Group (HLEG) in 2018 and 2019, in the attempt to draw up *Ethic Guidelines for Trustworthy Artificial Intelligence* (see https://ec.europa.eu/futurium/en/ai-alliance-consultation/guidelines#Top). Here it is:

> Artificial intelligence (AI) refers to systems that display intelligent behaviour by analysing their environment and taking actions – with some degree of autonomy – to achieve specific goals.
>
> AI-based systems can be purely software-based, acting in the virtual world (e.g. voice assistants, image analysis software, search engines, speech and face recognition systems) or AI can be embedded in hardware devices (e.g. advanced robots, autonomous cars, drones or Internet of Things applications).[2]

To judge by such definition, there seems to be no major difference between the devices listed as examples in the definition above and human behaviours. You can actually see, on one hand, that emphasis is placed on the "behaviour" of intelligent entities, both human and artificial, and, on the other side, that the words used for such behaviours are *the same* in either case. After all, if we take a behavioural approach, such consistent, standardising use of vocabulary is certainly not a problem. But, as I have shown, this is a premise that is implicitly and noncritically accepted.

This anthropomorphising process is certainly understandable. We are misled by the language. We have no other words to say what is different from us except those we have shaped on ourselves and on relational models that revolve around us. Here, *human-centered AI* is present but seems to have been bent into dangerous forms. But, again, something more extreme happens here. We slip into a veritable argumentative short-circuit, which we should be well aware of. First, we regard machines through the lens of our actions. Then, we end up reflecting ourselves in them, regarding them as models of our own behaviour.

I am going to sum up a number of passages that are implicitly taken here. For the lack of a better word, for the operation of some devices we use the same terms as we use for human beings: first and foremost, the term "intelligence". Machines are thought of as human beings. However, machines, in their procedures and their "actions", may be explained and reproduced, to a certain extent, in their hardware and software. Reproduction is thought of in terms of a programming procedure. Then, depending on the level of effectiveness, efficiency and straightforwardness of such model of explanation, reproduction, programming and control, it is applied to humans as well. And it can be, just because the words used in both cases are the same. In other words, a sort of retroaction—a definitely paradoxical one—happens whereby what we needed in metaphorical form to understand the operation of certain devices is now taken in the literal sense to explain the operation of the human being, that has now been identified with the device.

All this has multiple consequences. It is obvious that, if we attribute human features to what is not human, not only do we think of it as being like us—and in this case I have spoken of the "Disney effect" somewhere else—but, most importantly,

[2] *A Definition of AI: Main Capabilities and Disciplines*.

we are prepared to project our expectations and interests, our illusions and fears onto such entities or even go so far as to consider them better than us and therefore submit to them. But there's more than that, something about the conditions, rather than the consequences, of such approach. As I briefly mentioned, if a clear distinction between human and artificial agents disappears, no ethics may be possibly developed for AI. It cannot be possibly done because, if the difference between the two forms of action is disregarded, the possibility of a relation between them is eliminated, so investigating and discussing the proper, right, adequate ways—defined as such based on compelling criteria and principles—to build such relation becomes irrelevant.

2.3 The Ethical Framework

A well-grounded acceptance of such difference is therefore a prerequisite for an AI ethics. AI ethics is actually a specific form of relational ethics. The forms and ways in which such relationship takes place now have to be introduced and analysed. I will start with a brief overview. Some of its deeper aspects will be thoroughly investigated in the next paragraphs.[3]

Based on the perspective we have taken in this research, there are three ways in which we can speak of an AI ethics. Based on these ways, the approaches taken by the literature so far in addressing this issue may be incorporated and further developed. A distinction must be made between:

- An ethics embedded in AI-endowed programmes and machines;
- An ethics for the ways in which human beings interact with AI-endowed artificial agents;
- An ethics as an extension of shared ways of living in contexts that have been opened by human action, by the actions of artificial entities and by the interactions (between themselves and with the environment) between these two mechanisms of action (the human and the artificial one), respectively.

If we wanted to mention what we really need to investigate, in terms of our behaviour as human beings, as far as these three points are concerned, we would say that we would support:

- The "training" of machines and a specific design of the programmes that will embed ethical criteria, avoid bias and be ethically monitored;
- The "training" of humans in the ever-new contexts that the technological advances involve, in which humans can ethically interact with the action of artificial devices and with the environments they operate in;

[3] There are lots of foundational books on this topic. Let's just mention (Coeckelbergh, 2020), (Dubber et al., 2020), and S. Voeneky, Ph. Kellmeyer, O. Mueller, W. Burgard (Eds.), *The Cambridge Handbook of Responsible Artificial Intelligence*. Cambridge: Cambridge U.P.

– The search for globally shared (universal) values to handle global (transnational) problems, that is, a key clarification and a specific elaboration of what "ethics" means in such contexts.

2.3.1. Let's quickly look into these three forms in which ethics maybe possibly set, in connection with AI. The first approach—let me say that again—is the one that aims to programme and train, based on ethical criteria, the AI-endowed devices. In this case, I am speaking of an ethics *in* (i.e. enclosed, encapsulated—*embedded*—in) the AI. This is something that is processed and embedded in different ways (for example, in the design, in the programming and in the building of the technological devices) by humans.

But a problem arises in this respect, a problem of *control*, of human control, due (among other things) to the machine-learning techniques, and a problem with the (usually low) algorithmic transparency i.e. *explainability*. How can then such control be retained (providing it is possible and provided it is necessary)? This is the question that underlies the idea of a "*human-centered AI*". The answer which is usually given to that question is one that tries to impose *limits* on the action of machines: maybe even making the programme limit itself and resort to humans, at some time of its operation, to finish the job. We'll see that more extensively in §4.

The underlying problem of this approach is that it demands we go in the opposite direction to the mainstream vernacular (which, conversely, thinks humans should stretch beyond their limits with the help of machines and programmes: see (Coeckelbergh, 2020 [Chap. 2])). This is about embedding limits in the machines themselves. This is an ethics that is incorporated, indeed "embedded", in the machines themselves. But to develop it—let me say that again—the idea that machines can only help upgrade human action must be challenged. We must change the paradigm: from AI as a *replacement* for humans to AI as a *complement* to humans, in cooperative ways.

2.3.2. The second form of ethics is instead the one that takes over the task of regulating human action in its relationships with the actions of AI-endowed devices. Here, it is the human being that stands in front of their developments, assesses them, judges them. In this case, I am speaking of an ethics *of* AI, i.e. that concerns the conditions and consequences of such dimension. It is an ethics that comprehensively deals with the interaction between human action and artificial action. It investigates, justifies and regulates the ways in which humans adapt to and put themselves forward in new situations, even go so far as to change their behaviours, more or less deliberately. The above pages have been an example of this approach.

2.3.3. In these cases, though, a question arises. The question is: what are the values we look for, for the development of such ethics? What are, in other words, the values based on which both machines and humans *would do well to* act? To answer these questions, many are the approaches that can be taken. Usually, a principlist approach is taken (Floridi et al., 2018). In the following pages, instead, the chosen approach is a bottom-up method. We will start by taking the current ethical and deontological codes and will compare them. We will ask ourselves whether there are any shared values, and which ones they are.

The aim of such search is to achieve a preliminary snapshot of the values at stake. They are, or should be, universal values, because we resort to them to deal with global problems, though always in the contexts they have been used in. At any rate, they are values that require to be further justified in ethical terms, without sticking to a purely legal approach, which more often than not turns out to be ineffective. They are mainly values that overshadow other values, which could perhaps better cope with the problems that the approach at hand leaves still open. Therefore, they must be defined and analysed, and those most suitable for our purposes should be looked for.

3 Common Principles and Shared Values for an AI Ethics

Let us focus on the search for common principles and shared values that can provide a common ground to build an ethics of AI upon. To do so, we will analyze the landscape of AI codes of ethics and guidelines that have been drafted in recent years (Boddington, 2017; Laarson, 2020).

Especially since 2016, there has been a proliferation of these documents and of the principles they contain. In AI ethics, these initiatives were welcomed as a positive sign of growing involvement by the international community. However, especially in the case of private companies, they were seen as attempts to avoid stricter forms of control and improve their brand reputation through some form of *ethics washing* (Wagner, 2018; Floridi & Cowls, 2019; Metzinger, 2019) or «cosm-ethics» (Goffi, 2019). Although this risk is real, it should not be used as an excuse to play down the importance of ethics (Bietti, 2021); instead, it should act as a drive to investigate the global landscape of these documents and improve their effectiveness. Codes of ethics and guidelines share a principlist approach (Mittelstadt, 2019; Seger, 2022), typically used in applied ethics (Frey & Wellman, 2003; Fabris, 2018). (Beauchamp & Childress, 2019) have proposed this approach in bioethics. Notably, four principles are considered in this field: autonomy, beneficence, non-maleficence, and justice.

Principles are a valuable part of applied ethics, as they help condense complex ethical issues into a few core elements and form a common ethical ground, with shared values. They guide institutional and professional decision-making processes and help address public concerns by clarifying the industry's ethical commitments. Therefore, the ethical significance of these documents should be investigated by analysing the currently shared principles and by promoting some new aspects that have been left out of the current debate. In the face of this proliferation of documents and principles, many scholars have compared documents and identified shared core values. Here, we will review the debate on such issue, highlighting how difficult it can be to find some widely shared ground, if not a universal one. As a solution, we will then propose to resort to different, more relational principles, which have been overshadowed so far in this landscape of documents. In particular, we will focus on the concept of vulnerability and its ethical potential.

3.1 The Non-existent Universality of Current Reference Values

To begin with, a reflection on the use of the word "universal" is in order. In the history of thought, such concept has been lengthily and widely debated (Armstrong, 1989; Di Bella & Schmaltz, 2017; Galluzzo & Loux, 2015). So, it is important to explain what it means when used in this specific context. For the sake of brevity, we can allude to the idea of participation (Pap, 1943): what is shared by all and which, therefore, all participate in, by analogy or through some relation of commonality, is universal. Not only that, what is directed to everyone, and therefore what everyone can identify with, is universal too. Let's think of the Universal Declaration of Human Rights (1948): it applies to the human community, regardless of race, gender, religion, or physical appearance. In addition, it aspires to express principles and values which everyone, regardless of their cultural background or individual sensitivity, can adhere to. So, at the same time, it means something that is addressed to everyone and is acknowledged by everyone.

As to the universality of codes and guidelines, then the treatment of the topic of universality should be taken in a double sense: first in terms of the *convergence* of the existing codes, secondly in terms of the *shareability* of such codes from a pluralist perspective. To put it differently, on one hand we try to find agreement and communality between the principles laid down by the codes, so as to provide a common synthesis; on the other hand, instead, we ask ourselves whether such perspective can be representative of the different personal, and above all cultural, sensitivities as a whole.

The following meta-analysis compares various sets of documents to incorporate AI Principles into a consistent framework. Other meta-analytical work, not accurately mentioned, has been done in AI ethics (Schiff et al., 2022; Loi et al., 2020; Wright et al., 2020; Rudschies et al., 2020; Attard-Frost et al., 2022). However, these six documents were selected for specific reasons. (Jobin et al., 2019) is the first systematic study and one of the most widely cited. Although (Floridi & Cowls, 2019) and (Hagendorff, 2020) focus on a limited number of papers, they offer a solid interpretative perspective, which helps elaborate a more structured AI ethics. Compared to the other studies, (Zeng et al., 2019) and (Fjeld et al., 2020) examine principles through a more in-depth semantic analysis, based on keywords from general topics with comparable meanings. This makes it possible to delve deeper into the themes of the principles, beyond their synthetic enunciation. (Correa et al., 2023) is the most recent study that examines the largest number of documents. It compares not only documents but other comparative studies as well. These aspects make it valuable and original.

- The scoping review by (Jobin et al., 2019) examines 86 documents containing ethical principles or guidelines. These documents were drafted from 2011 to 2019 by private companies, governmental agencies, academic institutions, intergovernmental or supra-national organisations, non-profit organisations and professional associations/scientific societies, private sector alliances, research alliances, science foundations, federations of worker unions, and political parties.

They notice a convergence of these documents around 11 ethical principles, namely transparency, justice/fairness, non-maleficence, accountability, privacy, beneficence, freedom/autonomy, trust, dignity, sustainability, and solidarity. Of these, the first five are those that appear most frequently in most codes.
- In their study, (Zeng et al., 2019) analysed 27 documents, grouped by background: Academia (including non-profit and nongovernmental organisations), Governments, and Industry. They elaborated a semantic analysis and identified ten general topics: humanity, collaboration, sharing, fairness, transparency, privacy, security, safety, accountability, and AGI/ASI. Every topic contains semantically equivalent keywords.
- The meta-analysis proposed by (Floridi & Cowls, 2019) takes into consideration six high-profile initiatives established in the interest of socially beneficial AI, selected as recent, relevant and reputable: the (Asilomar AI Principles, 2017), the Montréal Declaration for Responsible AI (University of Montréal, 2017), (IEEE, 2017), (EGE, 2018), AIUK (House of Lords Artificial Intelligence Committee, 2018), (The Partnership on AI, 2016). They summarise the 47 principles in these codes in an overarching ethical framework, based on the four classic principles of bioethics (autonomy, beneficence, non-maleficence, and justice), adding explainability as the new enabling principle of AI.
- (Fjeld et al., 2020) found eight common themes from a dataset of 36 documents: privacy, accountability, safety and security, transparency and explainability, fairness and non-discrimination, human control over technology, professional responsibility, and promotion of human values. They analyse these principles in terms of stakeholders, content, geography, and date.
- (Hagendorff, 2020) analyses 22 of the most significant AI ethics guidelines and issues recommendations to identify the overlapping principles and divergences in the different codes, so as to propose forms of implementation and improvement of AI ethics. The principles addressed in most of the documents (over 80%) are accountability, privacy, and fairness, but safety/cybersecurity and transparency are also frequently mentioned.
- Finally, the most recent study, covering the highest number of documents, is that of (Correa et al., 2023). The dataset they examined consisted of 200 codes of ethics and guidelines, and 17 principles were identified: accountability/liability, beneficence/non-maleficence, rights of children and adolescents; dignity/human rights, diversity/inclusion/pluralism/accessibility, freedom/autonomy/democratic values/technological sovereignty, human development/education, human-centeredness/alignment; intellectual property, justice/equity/fairness/non-discrimination, labour rights, cooperation/fair competition/open source, Privacy; reliability/safety/security/trustworthiness, sustainability, transparency/explainability/auditability, truthfulness.

A comparison of such studies detects some convergence around a few principles, i.e. transparency, privacy, accountability/responsibility, fairness/justice, safety/security, freedom/autonomy.

- *Transparency.* This principle[4] is the one that features most often in the codes and in the debate around AI ethics (Larsson & Heintz, 2020). It includes both the transparency of the system and the transparency of the organisations that use it. So, this concept applies as much to the technological as to the human side of the matter. Actually, more and more often the class of deep learning-based AI systems involved in the decision-making processes seem opaque and provide results that are unintelligible to humans (the so-called "black-box models") (Guidotti et al., 2018; von Eschenbach, 2021). Transparency then is a key factor to encourage trust in the predictions, recommendations and the ensuing decisions of the algorithmic systems (European Commission, 2019). At the same time, the organisations that use such systems must therefore guarantee transparency about their purposes and the potential risks associated with the use of such systems (IBM, 2017). This is also a way to protect those who have been damaged by an AI system and challenge the results of predictions, recommendations or decisions based on simple, easily-understood information (OECD, 2019). Within most documents, transparency is associated with the concept of explainability (Preece, 2018).
- *Privacy.* The use of large amounts of data as required for the operation of these systems justifies the central role of this principle in documents. It is closely related to data protection standards outlined in the General Data Protection Regulation (GDPR) (UK Government, 2018). The ethical guidelines of the organisations focus on protection and the prevention of non-ethical uses of personal or sensitive information. They also impose anonymisation procedures. Personal information must remain under the users' control, and users must be certain that their personal information is processed for lawful, legitimate purposes, without limiting their personal freedom (Asilomar AI Principles, 2017).
- *Responsibility/Accountability.* Lots of responsibility and accountability issues arise from the use of autonomous and semi-autonomous AI systems (Santoni de Sio & Mecacci, 2021). This principle is related to both chargeability and answerability. (Novelli et al., 2023) Even when an AI system is opaque, the institutions and organisations that use it must be accountable for the decisions that derive from such use (ACM, 2017) and must answer for them with the authorities and with society. Responsibility also involves an honest behaviour on the part of developers and organisers, who must disclose any mistake or security risk (University of Montreal, 2017). Monitoring the processes, recording the operations and keeping registries are all necessary ways to demonstrate who has been responsible for what (IEEE, 2019).
- *Justice/Fairness.* The predictive potentials of algorithms have boosted the use of AI systems in many areas, including insurance and financial markets, personnel selection services, and the judiciary. However, such applications have raised

[4] (Turilli & Floridi, 2009) claim that this is not a veritable principle but a «pro-ethical condition for enabling or impairing other ethical practices or principles». Without going into the details of whether such assumption holds true or not, we will just consider it a principle, as it features in many codes and guidelines in these terms.

problems by creating inequalities, discrimination and exclusion (O'Neil, 2016). The wide scope of concepts of justice and fairness (Giovanola & Tiribelli, 2022) tries to gather all these problematic issues into one single principle. The documents offer a negative and a positive meaning of justice. On one hand, they draw attention to the need to limit the biases (whether based on race or gender or anything else) inherent in the data (FAT/ML, 2016) and to avoid mistakes that could lead to unfair answers (Telefonica, 2018). So, they are concerned with avoiding discrimination and inequality. On the other hand, they also highlight the importance of promoting equality and equal opportunities for everyone (European Commission, 2019; Atomium European Institute, 2018).

- *Safety/Security.* Generally, these concepts are relevant to cybersecurity and the potential vulnerability of the systems to deliberate attacks. To avoid malicious uses and manipulation, which may result in breaches of privacy[5] and other damages, secure-by-design systems are recommended (Telefonica, 2018). Even after the production process, the systems need to undergo regular risk assessment procedures. Promoting inspections and tests on the systems' conditions and robustness throughout their lifecycle is a priority (Asilomar AI Principles, 2017).
- *Autonomy/Freedom.* AI systems may impose constraints on decision-making and freedom of choice. Because of this, the documents put the principle of respect for autonomy and self-determination at the centre. Such concept is addressed insofar as relevant to consent, that is, being aware of the way the data are processed (Council of Europe, 2017). Since machines have the ability to take over part of human tasks, in this context autonomy also consists in the power to choose when a decision should or should not be handed over (Atomium European Institute, 2018).

  Extending human autonomy is often associated with control over (and therefore with restricting the autonomy of) the machines (University of Montreal, 2017).

Before accepting this convergence around these principles from the codes as an obvious sign of the sharing and universality of values, it should be pointed out that many differences lurk behind the many similarities. Indeed, despite finding quite a lot of principles in many codes, no common principle or value could ever be found in these texts (Jobin et al., 2019; Hickock, 2021). This is partly due to the fact that the same concept can be expressed differently, in different terms, and the fact that most of these principles overlap in some parts of their domain, so they are classed differently depending on the context. However, even when the same term is used, the way a principle is construed in each document can often differ widely (Ryan, 2020).

The type of issuing authority is particularly influential on the role and presence of each principle (Rudschies et al., 2020; Schiff et al., 2021). In particular, in the codes issued by governmental bodies, greater focus is given to matters such as *privacy* and security, but there is little interest in *accountability;* the latter takes centre

[5] The problem of security is often associated with that of privacy (RenAIssance Foundation, 2020; Telefonica, 2018; Microsoft, 2018).

stage instead in the texts drawn up by academic boards, while for companies the issue of cooperation between stakeholders takes centre stage, at the expense of privacy issues (Zeng et al., 2019). Such results are therefore suggestive of remarkable differences even in the allocation of responsibilities, the experts' role in the development of the codes, the inclusion of public members, and the relationship between economic interests and values. This means that the alleged convergence between different codes hides many divergences as well, so that the first meaning of universality cannot be fulfilled in these terms.

The second question about the universality of the codes concerns the field of moral pluralism, both individually, as we are all moral agents who have our own view of the world, and culturally. It is especially the cultural plane that is a critical point. Cultural and religious differences actually affect the understanding of principles and values in fundamental ways, and this challenges the achievement of a single ethical framework. Some principles can be the very embodiment of a culture and be almost non-existent in another: for instance, the Confucianism-inspired Chinese perspective puts society at the centre as a subject of interest, while the North-American one, rooted in Protestantism, is more focussed on the individual (Hine & Floridi, 2022). Even in the concept of privacy, it has been found that there are key differences between the Western perspective and that of other regions of the world: in India for example social life is organised without thinking of the protection of personal privacy as an absolute value (Marda & Acharya, 2014), while in the Western perspective the ownership of data and information is a central issue and an extension of the very idea of identity.

Most of the codes and documents reviewed are developed within the More Economically Developed Countries (MEDCs) (Jobin et al., 2019), i.e. Europe, North America and China. The other continents, the so-called "Global South", are underrepresented, and the ethical works proposed therein are not properly represented. For example, Africa has its own thoughts on AI ethics at a political-institutional level and at an academic level (Gwagwa et al., 2020, 2021), mainly based on the Ubuntu philosophy (Mhlambi, 2020; Van Norren, 2023). In South-America too, lots of efforts are being made to establish a standard ethical framework for AI systems (Mancilla-Caceres & Estrada-Villalta, 2020), but such considerations are considered negligible and hardly ever taken seriously. Even approaches inspired by an alleged greater universality, such as the one based on human rights (Wong, 2020; Liu & Lin, 2020), often elude such difficulty in attempting to reconcile and, above all, merge different views, to such an extent that there has been talk of an «algorithmic coloniality» (Mohamed et al., 2020). So, this entails an idea of universality that is deeply distorted by the bias of the perspective at hand. As (Goffi et al., 2021 [46]) claims, «such behaviour is clearly the product of a biased stance, mostly the idea that some countries hold an indisputable truth regarding what is acceptable and what is not». By stating these principles and values and by presenting them as universal, there is the risk of leaving equally good perspectives out of the debate as irrelevant. As stated by Kiemde and Kora (Kiemde & Kora, 2022): «the predominance of Western input on AI ethics guidelines can lead to a dominance of Western values and vision on AI ethics».

3.2 *The Need to Resort to Other Principles*

Faced with such issues, we can conclude that the principles contained in the current codes of ethics, though highlighting some fundamental needs in the context of an AI ethics, are not a universal compass, either in the convergence of the different documents or in the global shareability in a pluralist moral context. So, this is not about considering the codes a failure or dismissing the principles they advocate, this is about understanding whether such principles are the only possible ones for a shared horizon or whether more can be identified and added to the picture.

In a nutshell, we must ask ourselves whether such universality can be reshaped around new principles. (Hagendorff, 2020) points out that the principles contained in over 80% of the codes are abstract, can be calculated and mathematised. Regarded in mathematical terms, they can be implemented through technical solutions and are based on an individualistic, abstract notion of interhuman relationships, leaving out the social contexts and power dynamics. Others are instead underrepresented, if not completely missing.

Such omitted principles recall the so-called «forgotten issues of AI ethics» (Hagendorff, 2022): the use of AI for political propaganda, the risk of social fragmentation, the ecological footprint, the poor working conditions of clickworkers, and diversity in research and development (Crawford, 2021). To deal with such kind of problems, we must come up with some principles of the relational dimension, and look not only at the stakeholders but also at the contexts in which a relationship with a system develops: namely, the social and economic context, the environment, and generally the network of relations that enwrap us in our interactions with the systems.

The corresponding principles are the fundamental values of our society and are at the heart of several constitutional charters worldwide. Terms like "care", "solidarity and support" (Luengo-Oroz, 2019), "welfare", "social responsibility", "sustainability and environmental responsibility" (Roche et al., 2021; Genovesi & Mönig, 2022)[6]; the so-called "Principles of DEI", namely "diversity", "equity", and "inclusion" (Cachat-Rosset & Klarsfeld, 2023), hardly ever find a place in AI codes. Even when they do, they are vaguely conceptualised and, for this reason, poorly operational.[7] Hagendorff argues that the most prevalent principles in the AI guidelines coincide with the morality of justice, while the neglected ones reflect the values of care (Gilligan, 1982).

---

[6] «When we released the draft data visualisation for feedback, we were frequently asked why sustainability and environmental responsibility did not appear more prominently. While the authors are sensitive to the significant impact AI has and will have on the environment,18 we did not find a concentration of related concepts in this area that would rise to the level of a theme […]. As the conversation around AI principles continues to evolve, sustainability may become a more prominent theme» (Fjeld et al., 2020).

[7] The AI Now report (Crawford et al., 2019) is an exception. In this document, issues of diversity and risks for democracy and sociality are extensively considered.

What can we learn by looking at these glaring omissions? We feel that the introduction of such issues and the addition of these more relational principles could be a good option to create a common ethical ground.

In this sense, a principle that is almost totally missing in the documents, that is, vulnerability, deserves special attention. At the centre of the philosophical, bioethical and political debate (Ten Have, 2016; Ferrarese, 2017; Gilson, 2014), vulnerability describes both the common condition of all living beings and the specific situation of exposure in which some individuals find themselves living in. This notion is hardly ever found in the codes, and never as a principle. For example, in the HLEGAI (European Commission, 2019), vulnerability is used two ways: it stands for vulnerable groups, such as disability and childhood, but it also stands for the vulnerability of systems. It is used only for specific categories and ordinarily understood as opposed to the idea of autonomy as a standard condition of human beings.

However, vulnerability is not an exception, it is an existential condition of finitude and fragility, which requires a caring approach and a sharing of responsibility with all humans. Even in their relationship with technology, humans are exposed to new risks and turn out to be vulnerable, then, in many respects (Coeckelbergh, 2013). (Helberger et al., 2022) describe the concept of «digital vulnerability» by claiming that, in the digital arena, we are all exposed to some form of vulnerability, in architectural and relational terms, where «architectural» means that it is not fortuitous but that such vulnerability is exploited by marketing and nudge strategies; and where «relational» means that it is inherent in the asymmetrical relations of power and knowledge.

(Malgieri & Jędrzej, 2020) speak of «vulnerable data subjects», meaning the vulnerability of data processing and its results. In the former case, it is inherent in the limitedness of free consent to the sharing of data, the difficulty in understanding information or assert one's rights in data protection. With the results, instead, data processing can damage the data subjects and lead to some form of discrimination, not to mention physical and psychological injuries.

Such term is also used to speak of the vulnerability of systems, not just human vulnerability. It is a state of weakness in the software or hardware or the lack of security procedures or of technical, physical or other types of control that could be exploited by a threat (Pipkin, 2000). Actually, digital systems are exposed to attacks, breaches and attempted tampering for purposes other than those they have been programmed for. In this case, the resulting risk is, of course, dependent on human will (often the will to do harm), but it is also dependent on the lack of technical protection against such kind of sabotage (Kizza, 2009).

The proposal we are putting forward herein, to highlight this principle in the ethical framework of AI, is mainly because it helps give a more substantial content to the concept of dignity. This substantiating value makes it possible for any human being, no matter how autonomous, to be involved. It actually brings together a universal meaning (we are all vulnerable) and a particular meaning (everyone is vulnerable to a different extent and in different contexts) and relates them to the specific vulnerability of systems. In addition to a descriptive role, it also has a strictly relational character: the protection of vulnerability actually recalls care and responsibility (Dadà, 2022). This principle may help rethink the interaction between humans and systems in their relational sense, while finding the contexts in which some forms of discrimination and inequality happen and outlining an allocation of responsibilities.

Therefore, the principle of vulnerability mainly pursues the overall protection of the human being and his relationships, as well as highlighting each event, each specific situation in which such vulnerability is amplified by economic, political or technical choices or, as in the case of the AI systems, by technical choices in their coding or development. If all human beings must be protected from the potential damage caused by the use of AI, each one is specifically protected, depending on the context and conditions. The more vulnerable ones must then get additional protection from the potential damage caused by the AI systems. The vulnerability of systems affects the vulnerability of individuals and needs therefore to be addressed. So, it seems to be a useful litmus test to identify situations of discrimination, health and safety hazards, and the sharing of responsibilities (individual, political, social) in digital environments.

We can state therefore that the search for universal principles and common values is not so easy, and the current codes provide only part of the answer. If we want to be able to build a shared ground of universal values for an AI ethics, taking the principles currently laid down by the codes is not enough: universality as convergence misses the mark, then, in its efforts to build a common ground. Paradoxically, it points us to some common omissions that have been left out of the current debate, that need to be filled. The search for universality is a work that cannot boil down to a synthesis of what it is; it should be a shared and participatory building work.

Referring to the omitted principles such as vulnerability is one of the cornerstones of this ethical effort. Such principles are also a space for potential shareability, beyond the boundaries of a Euro-centric perspective. For instance, the African Ubuntu philosophy offers an idea of individuality that is rooted in an interdependence with the community and with the environment, with care for the common good and the protection of fragilities within a collective perspective. Such perspective has been regarded by some African scholars as a potential starting point for a post-colonial AI ethics and is perfectly consistent with such horizon. As opposed to the Western individualist perspective based on competition, the Ubuntu theory is based on the relational idea that "*I am a person through other persons*" (Van Norren, 2023). Such perspectives can help expand the horizon of the shared principles of AI.

Such principles call to mind, though in ways that are harder to mathematise, a collective effort to achieve an ethical ecosystem for AI. So, in the domain of the current ethical codes, they look more suitable for enhancing those features that have been pushed into the background and that could pave the way to a more widely shared perspective.

4 *The Endless Tuning*: How to Pair Different Kinds of Agents

4.1 In the Beginning

The principle of vulnerability can deeply interweave the ethics of artificial intelligence. An ethics of relations *with* and *in* artificial-intelligence devices is ambivalent: on one side, it is a form of ethics among many others, such as utilitarianism, deontology etc., on the other side, it could lead to a metamorphosis of ethics as such. Here, "relationship" is taken to mean a *beginning* more than a *principle* (Buber, 1970, 72; Fabris, 2016b [172]). It's no coincidence that every ethical principle mentioned so far hints at some relationship—let's mention respect for the user's autonomy or the commitment to answer for the action of a device in a court of law.

If technology brings in something new, it is that … «and yet it moves». In the cyberworld, then, a value is not only pursued, but also achieved in real time. The bottom-up perspective, the empirical search carried out so far, and the evidence of the geographical and cultural boundaries of the ethical codes on artificial intelligence suggest that what matters is not so much the universality of the values at stake but their universalizability (Fabris, 2016b [167]). A relational approach to ethics—in this case, the ethics of artificial intelligence—does bond thought and praxis together, because a relationship is a practice driven by an effort: first, the effort to keep in touch, to continue the dialogue. Such an approach can bring the high and the low together: it tries to be virtuous but it is not individual, it pays heed to the consequences of its actions but it knows that it cannot perfectly master them, it keeps its duties in mind but it is also mindful of the peculiarities of the other entities—in this case, the machines—that have to be dealt with.

Now, extending principlist bioethics to artificial intelligence through a fifth principle of explicability (Floridi, 2023) is restrictive, if for no other reason that an intelligent device is not a passive entity like a molecule, nor is it perfectly clear who is the doctor and who is the patient (Atomium European Institute, 2018). Nevertheless, we could learn a useful lesson from bioethics—or better from an event that concerns it. As to the drafting of the Belmont Report, which was promoted in 1979 by the U. S. Department of Health and Human Services and laid the foundations of the four principles of autonomy, beneficence, non-maleficence and justice, Stephen Toulmin told:

> […] when you look at the origin of those principles, it's very interesting to notice that there was much more agreement between the different Commissioners about the kinds of research which were morally acceptable or unacceptable than there was about the ways in which the reasons for their consensus were to be justified. At the end of each period of discussion, […] they would find that of the 11 Commissioners, there were perhaps two who disagreed with the general tenor of the rest of the Commission. But the ways in which they expressed the reasons for their agreement or disagreement were very different from the formulation of the terms on which they were very largely agreed. In fact, one might say that the Protestants agreed for Protestant reasons, the Catholics agreed for Catholic reasons, the Jews for Jewish reasons, and so on (Schwetz & Toulmin, 2004).

Agreement prevailed, and the principles took off from there. So, the principles were the weighted average of the baggage of values of each Commissioner, weighted on a case-by-case basis. Carrying this into the province of an ethics of artificial intelligence means designing intelligent machines that we can relate *to,* without being seduced by the aesthetics of the user interface, and that are moral, not because they are artificially conscious, or even less because they simply embed ethical constraints into the source code (it would be impossible to predict all the variables *in the wild*), but because they foster relationships between people, each one adapted to its cultural preferences, by laying the grounds to make them happen. In other words, devices, the use of which—continuing the comparison with the ups and downs of the Belmont Report—may act as a real-life case for the application of some values and the emergence of some principles. Because it is always and only up to humans to lay down the purposes of action and undertake responsibility for them.

Then, a serious problem lies in the very concept of design: it should be a sort of pro-ethical design (Floridi, 2014 [190]), one that encourages the user's freedom. Yet, it is always "things", plain objects, that are designed. Instead, despite not being actual subjects, not only are intelligent devices ever more autonomous, but their autonomy, if meant as a form of (partially) controlled "independence-from" (see the American Society of Automotive Engineers' taxonomy of autonomous vehicles (Robertson et al., 2019)) looks fragile: in terms of actual robustness—resilience to unexpected inputs—and with respect to some distinctively atypical and unexpected behaviours. There are at least three cross-sectional clues to unpredictability, which sound like a deceitful experiment of the hard ontology (objects, tools, products) attributed to artificial intelligent systems—indeed leaving, on the way back, a few raw nerves in our self-perception. Apart from the perspective of replacement and enhancement, as we are going to see, artificial intelligence could be meant as a fellow traveller.

4.2 First Clue to Unpredictability: Adversarial Attacks as Bugs in the Artificial Intelligence's Nature

How can one possibly make gross yet extremely accurate mistakes? It could be a special kind of mistakes, implicit in something's nature. (Descartes, 2008 [63]) noted that, for example, dropsy causes a deep feeling of dryness in the throat, countering «its usual cause, which is that the body's health would be benefited from drinking». Descartes tacitly glossed over it (better it happens rarely than often, «and the same applies elsewhere»), but he himself admitted that it was a bug in our nature. The intuition behind this, in a mathematically oriented world, was dreadfully revealing: a perfect God had made a mistake. Well, adversarial examples—and, conversely, the ML models' reactions thereto—seem to have something in common with Descartes' "dropsy" bugs. *Mutatis mutandis*, they might perhaps be bugs in the nature of artificial intelligence: it is like they appear not because models do not work properly, but because on the contrary they work too well. Indeed, (Szegedy et al., 2013) found out some instability in the neural networks w. r. t. small perturbations to their input, whereas they were supposed to be robust and capable of local generalization, i.e. it was expected that «for a small enough radius $\varepsilon > 0$ in the vicinity of a given training input x, an $x + r$ satisfying $||r|| < \varepsilon$ will get assigned a high probability of the correct class by the model». This assumption did not hold. The authors, indeed, set up some data perturbation r such that $f(x + r) = label$ (that is to say, if x is an umbrella, $x + r$ will *in fact* be an umbrella too), but approximated a minimum distortion function, so as to produce some input instances x' as extremely similar to x that *instead* had $f(x') \neq label$ (that is: a perfectly common umbrella is misclassified as a school bus or a panda bear). They called x' the "adversarial examples". An accuracy of 0% or little more was measured for different datasets, alongside cross-model and cross training-set generalization properties, suggesting that «adversarial examples are somewhat universal». They still are unharnessed, and extremely dangerous too: if a war drone detects a school bus as a dog or something else that is not included in a list of protected targets, it will not hesitate to throw a bomb in its surroundings (Schiaffonati & Tamburrini, 2021).

4.3 Second Clue to Unpredictability: Emergent Abilities of Large Language Models

(Searle, 1980) invented a "Chinese room"-like thought experiment, demonstrating that «instantiating a computer program is never by itself a sufficient condition for intentionality», being indeed just a manipulation of formal symbols. However, what if the Chinese learner of the experiment, only receiving and processing syntactical instructions, turned out to be capable of uttering some sentences, despite this not being explicitly mentioned in the instructions? This seems to happen with the so-called emergent capabilities of Large Language Models or LLMs (taken here as examples of the so-called Foundation Models). Despite, perhaps, not *being* minds, just like Searle argued, yet we are surely pushed to investigate why those anomalies happen.

LLMs are Deep Learning networks, able to vastly acquire natural language knowledge especially from the Internet, apparently handling the sense of words, profiting from reinforcement learning techniques and from the structure of transformers, fine-tuning on single tasks (Attardi, 2023). The abilities to which we are referring are deemed "emergent" (Wei et al., 2022) because the performance of the model with respect to them increases well-above random, only scaling beyond a *not precise* complexity threshold, measured f. e. by the quantity of parameters or the number of training FLOPs (floating-point operations). Neither do we know exactly why these abilities emerge, nor can we predict what they will be. (Wei et al., 2022) point to abilities such as few-shot prompting and augmented prompting strategies. Few-shot prompting means that a model is given a (natural language) prompt about a task, and it completes a response without any further training. Above 10^{22} FLOPs, some models proved themselves capable of, among other things, transliterating from the International Phonetic Alphabet and mapping conceptual domains. Augmented prompting means "extracting" abilities, as it were, through a maieutic method. Step-by-step prompting (chain-of-thought), for example, literally makes the model learn how to reason or follow instructions. (Webb et al., 2023) speak of "analogical reasoning", with GPT-3, on average, outperforming humans (with similar performance patterns) in Digit Matrix problems where rules had to be found out, such as transformations (distribution-of-3, progression etc.) and logic problems (with OR, AND, XOR). How to trace back responsibilities if even developers do not know what abilities might emerge, especially when hallucinations come out?

4.4 Third Clue to Unpredictability: Instability and Occam's Razor Limits in Explainable Artificial Intelligence

Let us call a model "transparent" when its parameters are knowable, though this alone is no guarantee of intelligibility. Let us call it "interpretable" or "explainable" if a *senseful* path is traceable from an input instance x to an output y. Interpretability of machine learning models is ill-defined (Lipton, 2016), perhaps due to «incompleteness in the problem formalization» (Doshi-Velez & Kim, 2017): it is not very clear what an explanation effectively is and to whom it should be directed.

Nonetheless, research in explainable Artificial Intelligence (XAI) aims at explaining, controlling, improving the accuracy and fairness of black boxes, or discovering new facts hidden in the dataset (Adadi & Berrada, 2018). While XAI is a main tool to pry into black boxes, it nonetheless suffers from some flaws. (Holzinger et al., 2019) claimed the necessity to learn causality, not just correlations, among representations—what XAI cannot actually provide, even if this happens for experts too, in a different sense. Here comes the point: an explanation is *meant to be true*. Instead, it appears we are getting swallowed by complexity. Machine learning models have become nowadays inscrutable like the world itself, which they indeed analyse through data, reproposing to XAI two deeply-woven problems pointed out—actually, regarding statistics—by (Breiman, 2001).

Firstly, the *multiplicity of good models*, which paves the way for accepting the *more convincing* explanation instead of the true one. Paraphrasing Breiman, there will be several explanations whose fit is acceptable, but «there is no way, among the yes-no methods for gauging fit». Stability, a quantitative XAI evaluation metric which validates if similar instances obtain similar explanations, grasps our attention. Sometimes, indeed, not only is there incoherence among the outcomes of different local explainers on the same instance, but it may also happen that multiple explanations are made available by a *single* local explainer for very similar processed instances, producing a kaleidoscope effect. This phenomenon is known as the *instability* of XAI models, and it seems to be a serious difficulty. (Razmi et al.,

2023) report a glaring example thereof: interpretation attacks on sequential electronic health records. Some perturbed input instances $\tilde{x}$ were produced, capable of fooling an attention-based interpretation function Φ applied to a classifier model f, therefore leading to correct predictions which were nonetheless followed by purposely mistaken explanations. Secondly, *Occam's razor limitations*. Occam's razor literally stated: «entia non sunt multiplicanda praeter necessitatem», *de facto* pushing for simpler hypotheses. However, simplicity conflicts with the faithfulness of explanations, because of the so-called *curse of dimensionality*: if the number of attributes gets too large, the watermark gets unintelligible; if too small, the hypothesis diverges from reality.

4.5 Mirrors

It feels like being in front of an aporia, a boarded-up road that lets us steal just a glance under the cover of the black boxes—aware that the white boxes, i.e. transparent-*by-design* devices such as decision learning trees or linear models, are outperformed by deep learning. So, we need to get around this. Instead of merely trying to open the black boxes, of getting to know them—as Descartes would say—clearly and sharply, let's take a different, certainly windy, road, of which at least we can see the end.

The mode of action that will guide us like a lantern could well hark back to the Nicomachean Ethics (1103a): «Virtues, however, we acquire by first exercising them. The same is true with skills, since what we need to learn before doing, we learn by doing» (Aristotle Crisp, 2004). Will we be able to learn by doing together with artificial intelligence?

(Skitka et al., 1999) describe an experiment. Sitting at the joystick of a flight simulator, eighty students had to complete eight tracks each. A dial of the display was held by a virtual assistant, which would have alerted the users of any failure or error. The assistant was only 96% accurate. Nonetheless, the users "obeyed" its instructions 100% of the times in which its predictions were wrong. This phenomenon is known as "automation bias". How many of us, while at the wheel, have ended up taking a wrong turn at least once in a lifetime and maybe got blocked up in traffic, by dint of blindly listening to the navigator's algorithm? Margins of error exist even in the most valuable devices; but once we climb through the looking glass—just like Alice did—and model ourselves on the technology (Fabris, 2012), we lose the ability to detect them. The mirror, as we will find out, is a metaphor that can go a long way.

As it seems, in addition to the Oracle of Delphi, nowadays Socrates has his fans among computers: the Socratic machines (Punzi et al., 2023). Not that they help end-users give birth to geometric proofs, yet "they know that they do not know" and hand it over to the human decision-maker when they are not sure. Some fine points might have been overlooked, anyway: Socrates, just by knowing he did not know, considered himself the most erudite of them all.

Suppose, for instance, we have to complete a task of classifying some images of squares and circles into one of two groups, where each element belongs either to class 0 or to class 1. We decide to train a neural network in supervised mode, its last layer consisting of just two neurons, *n0* and *n1*. Both output a number between 0 and 1, but the sum of the two numbers is always 1, so that, to all intents and purposes, *n0* and *n1* produce two probability measures. Having taken an element as an input, if *n0* produces the highest value, it means that the element belongs to class 0 (squares), while if *n1* produces the highest result the element will be predicted to belong to class 1 (circles). The point is that there's a big difference between opting for class 0 with $n0 = 0.55$ or with $n0 = 0.99$, though the result is the same. In the latter case, the model has far greater confidence, in other words it is much more self-confident—but beware: being self-confident and being sure of what one says, i.e. in a broad sense *confidence* and *accuracy*, are not the same thing. If confidence drops below a given confidence threshold, the model throws up its hands and passes the buck to the user. That's why this technique is called *learning to reject*. While classification is the sorting of classes, rejection is the sorting of examples (Zhang et al., 2023). And if, on top of that, our model even tried to calculate—mind you: of its own accord—who (or what), between itself and the user, would be more reliable in dealing with a given input, letting the user or itself have the last word, it would implement a form of *adaptive rejection learning* (Madras et al., 2018) known as *learning to defer*. Shame, though: actually, the user is the only black box that is capable of discernment and that, being neotenous, can learn throughout its life. After all, could we ever believe that the model that knows it does not know and therefore is the most erudite of all would have really stepped back?

The flaw in the *human-in-the-loop* approach, *sic et simpliciter*, is that it plunges the user into a process without specifying which role it is given, and that's the reason why this wording could even contradict that of *human-centered artificial intelligence*. In the ethics of artificial intelligence, the artificial intelligence is not simply a mediator either (Benanti, 2022), a screen, a sometimes dangerously opaque one—here's the mirror, again—to reflect one's decisions in: because, as we saw, it is endowed with unique forms of autonomy. The cybernetics thinkers of the Forties used the behavioural method as an alternative to functional analysis, temporarily neglecting the specific structure of the object and its intrinsic organisation (Rosenblueth et al., 1943). This effective move set aside the constraints of ontological essentialism—the "what is" of devices—instead of crushing it altogether, so that on one side the essential difference—the metabolism (Ferraris, 2022)—between living beings and machines could be left in the background, and on the other side technological devices that could even be capable of phylogenetic learning i.e. reproduction of the species (Wiener, 1964 [13 ss.]) could be designed as if there were no tomorrow. That dream came true: *explainable Artificial Intelligence*, at least in the field of post-hoc global explanations, deals with white boxes that reproduce black boxes, which in turn try to get as close as possible to a set target. However, compared with that hard ontology that cybernetics thinkers had already dismissed, we are now seeing an unbridgeable swell. We are not speaking of some scary Golem, but of unfathomable computations, as in the architectures of the big transformers.

That's why, as we mentioned in Section one, if the *pars destruens* is ontological, then the *pars construens* will be ethical. There's the suspicion that the behaviour and the operation of artificial-intelligence devices should be regarded, ontologically, as being the same thing. It means that they do not look like objects or even less like subjects, they are "something else", relational beings *even though different from us*—and therefore with no personality or rights. Their autonomy, deeply observed, looks like the dark side of their opacity. If trust and control make up an antithesis, the key for a *human-centered approach* is building a bridge to communicate *with* artificial intelligence itself. In the tale of the Sorcerer's Apprentice—as recalled by Norbert Wiener—we could now read, not the rebellion of the neural networks but our inability to work out an effective language. Linear, one-way, instruction-based communication has fallen short of our expectations: maybe we should opt for a circular model.

We are full of fragilities, especially when we have to accurately grasp the details and take important decisions. If we cannot find the solution to an extremely simple equation straightaway, this is due to the laziness of what (Kahneman, 2011) used to call "System 2", the failed hero on the stage of our minds. If instead those who are reading these lines, maybe disagreeing with them or even getting bored, still cannot help to give a meaning to the signs printed on the white ground, let them blame the relentless lead of their thoughts, the "System 1" quickly taking the upper hand. You shouldn't trust those who jump the gun: System 1 is full of prejudices. The WYSIATI effect—*What You See Is All There Is*—is the emblem of narrow-mindedness; the *framing* effect, that makes us change our perception of a phenomenon depending on the way it is shown to us, is a gold mine for marketing; well known to sellers, the anchoring effect manages to snatch a little more money from customers when the negotiation starts at a higher price; the substitution bias instead makes System 1 step in when System 2 is struggling and answer the first problem available that is similar but easier.

System 1 is not such a bad character, though. It frames experience data into occasionally distorted stories, confuses the correlations with the causes, but it perhaps gives meaning and value to what it finds. This seems not to be the case with System 2, which in the progressive results of a golf player partly recognises the behaviour of a random variable that tends to regress to the mean. If this is the rationality of System 2, something must be wrong: freedom should not stumble upon the crazy maelstrom of fate. Maximising a utility function could be not enough to be called rational.

Maybe one could think that algorithms can make up for our shortcomings. After all, when we open up about our vulnerability, it means we are prepared to let ourselves be supported. However, if we still wanted to pair a human being to an artificial intelligence model, neither treating the former as a flawed System 1 and the latter as a corrective System 2 nor the opposite would be enough. Strictly speaking, the second assumption is not applicable, due to the limits of human cognitive faculties. As to the first one, artificial intelligence is anything but intellectually honest. A big chunk of bias—namely, «the inclination or prejudice of a decision made by an AI system which is for or against one person or group, especially in a way considered to be unfair» (Ntoutsi et al., 2020)—lies in data processing: in the low representativeness of the statistic samples or in the reckless use of (possibly) sensitive attributes, such as skin colour or postcode. However, such answers, even far too prompt—such as the above-mentioned *few-shot prompting* of LLMs –, give us an inkling that some distinctly algorithmic biases, the close relatives of the shortcomings of the human mind, do exist.

It truly seems that, similarly to our case, the two characters, "System 1" and "System 2", find a place in source codes too. Ironically, then, we should perhaps loosen our grip a bit and instead of controlling artificial intelligence systems—which is not always feasible—we should rule them, with two guarantees: doubt, which is an indicator of freedom, and competence, which is the fruit of studies and experience. If freedom needs to be preserved, ethics should not be procedural (Fabris, 2016b), but embodied, well-thought out, hard-fought and acted upon. Nevertheless, a computer is procedural by necessity. So, to pair man's freedom with the procedurality of a computer, we would need a kind of procedure that ceases to be merely procedural as soon as it is implemented. A kind of procedure, in short, that may create some *attunement*. Even the mirror must be able to reflect itself in us: that's how the mirrors become two.

A radio tuning circuit works in curious ways. The radio receiver generates a wave that is perfectly specular to the carrier wave coming from the antenna. As soon as they meet, one in front of the other, a beat starts between the two waves, which partly lessens them and partly amplifies them, until they are attuned. Here we are proposing a similar co-decision-making mechanism, in the sense of a conceptual alignment or mutual understanding (Stolk et al., 2016): setting two negative-feedback procedures one in front of the other—one mirroring the other, to go back to our metaphor again—as cybernetics teaches us. On one side, a human expert must be able to bind the learning and inference of an artificial intelligence device, correcting its mistakes, and on the other side the device must be able, when needed, to instil doubt into the user, so that they make the most of each other's vulnerabilities in a sort of cooperative game. This is about tuning the human being and the artificial intelligence, different kinds of agents, in a cyclical, infinity-shaped coming and going, constantly fine-tuning the abilities of both (Fig. 1).

I would just say in passing that here the concept of time takes on a peculiar meaning, which may be explained by another analogy. To show the intuition behind the restricted theory of relativity, Albert Einstein built a thought experiment, imagining an hourglass, made with two mirrors one in front of the other, marking time according to the coming and going of a constantly reflected ray of light. If the hourglass were moved forward (in a uniformly linear motion), the ray, following its movement and having to cover a longer distance diagonally, did not take longer: time changed its dimension itself (Balbi, 2021 [29 ss.]). In our case, of course, the constant is not the speed of light, it is the principle of relation, yet the time of relation—the learning and inference time of the hybrid system—almost seems to warp, as it has to match the freedom, limits, expectations and concerns of the human decision-maker with the constant flux of the environment (just think of a natural ecosystem or a disease

Fig. 1 A visual representation of the "endless tuning" loop. (Picture manually made by the author of this section)

of the body). Ditto for the concept of truth in the predictions and explanations, within the system itself, discovered, taken apart and rebuilt over and over again.

Without mentioning any specific architecture, let's see how such relational system could be structured. The artificial agent that assists us should have three main features:

1. For a specific task (such as looking at images), it must be pre-trained on a big training set when first supplied to an expert user and fine-tuned (for instance) on a lesser dataset, drawn from the history of user activity, in its ergonomics and environmental adaptation;
2. *Every* interaction must be recorded in a permanent log, for legal—tracking responsibilities—and technical reasons, except when data are removed due, for example, to privacy reasons;
3. The fine-tuning and inference processes must almost overlap, somehow shaping some form of *continual learning*, except when the entire pre-training must be done all over again after a major event.

If an expert user makes a new prediction of which he is sure, it will be up to him to update his own "tuning set", labelling the image he analysed as he deems most appropriate, and have the weights of his model updated accordingly. If instead he happened to receive a case he is unsure about and roped in an artificial intelligence model for help, the latter would compute a prediction, but it wouldn't disclose it straightway, it would rather build a sort of two-party dialogue in six steps, without tight time boundaries, to be strictly followed—provided one decides to follow them all—in this order:

(a) First, the expert user is asked to pass an opinion about the case, which is recorded;
(b) The expert user is presented with the answer, recorded in the tuning set or in the pre-training set, which was given by him or by those who built the pre-training set in a similar case. There could actually be inconsistencies;

(c) The user is asked if he wants to change his initial answer;
(d) To make the user reflect, according to the *machine-in-the-loop* approach (Miller, 2023), the model presents him with a number of facts on the case that could be explanations (e.g. saliency maps, but without predictions) or a neighbourhood of artificially-generated prototypes and counterfactuals (informing the user of the risk of producing unlikely images);
(e) The user is asked if he wants to change his initial/previous answer;
(f) The model tells its prediction to the user, providing a confidence level too, if possible.

At this point, going through the loop backwards the user will decide on the basis of his expertise whether to accept or reject the support provided. If he decides to reject it, he will be able to impose explicit additional constraints on learning (note, indeed, that updating the model's parameters on the tuning set is an implicit form of constraint) and, provided that the explainer is faithful w. r. t. the black box, on explainability—the model might actually have guessed the prediction right, but provided wrong reasons.

This coming-and-going, potentially unlimited process has been therefore called *the endless tuning*. It appears that such an idea would not counter fundamental rights, like the ones established in the Charter of Fundamental Rights of the European Union (European Union, 2012), particularly with respect to the right to personal integrity (art. 3), because concerns about the risk of undue influence on people's freedom are fully taken into consideration here. However, ultimately, who pays for damages, and *why*? (Chesterman, 2020) underlines that, while safety concerns arise around autonomous vehicles and deontological ones emerge around autonomous weapons, algorithmic decision-making «is tolerated provided that the legitimacy of such decisions can be ensured through the protection of rights and interests». It is not that legal tools do not exist. Instead, a clear concept of AI autonomy seems missing. For example, Italian law provides a legal instrument of objective liability, i.e. a kind of liability arising from the action itself, regardless of any subjective element of deliberate intention or negligence and spanning from the responsibility for a nuclear disaster to damages caused by animals or goods held in trust, related to the unpredictability and (un)avoidability of the damage (Bianca, 1995; Cian & Trabucchi, 1992). What is deeply interesting, especially in the latter case (see, for example, articles 2051–52 of the Italian Civil Code), is that the owner, or anyone else acting on the owner's behalf, is held responsible, unless it can be proved it was a fortuitous event; and all these examples resemble artificial intelligence in some respects—indeed, it is pervasive and powerful, it is a tool in someone's property or at temporary disposal, it resembles an animal in terms of autonomy. A most important function of the previously-suggested permanent log is precisely determining—for example by means of a judge—to what extent a possible damage was avoidable, and was it a fortuitous event or not, leaving aside the acknowledgement of effective intelligence for the time being. If it holds, issues about the allocation (and sharing) of responsibility might benefit from new assessment procedures.

5 Conclusion

In this paper, we have shown that AI ethics is a relational ethics, namely an ethics that investigates the different relational mechanisms that exist between a human and a system and is concerned with furthering the more equitable and fair ones, based on cooperation between the two agents. This is therefore a very wide domain, which can be investigated at different levels of reflection. Here, we have dwelt on three of such dimensions: on one hand, we have shown the key questions of the meta ethical discourse, in which the main concepts are outlined. From this perspective, the core question is about defining "artificial intelligence" and what "acting" means for a nonhuman entity. Having understood what "acting" means, we have lingered on the deontological perspective, the principles that encapsulate what "acting right" means. An analysis of ethical codes and guidelines allowed us to review the shared values within such framework. We pointed to the difficulty of achieving a truly universal perspective from such documents and the need to complete such overview by adding more relational principles that have been left out of the debate. In particular, vulnerability is a keyword. Such concept actually describes both the human's state of exposure to AI ethics (and the need for protection) and the situation of the system, which is also exposed to manipulation and damage. Prompted by such reflections, in Part three we have considered the most *embedded* dimension of AI ethics, i.e. how to design "intelligent" machines in keeping with relational ethics or, to put it another way, how to implement artificial-intelligence models that are structurally suited to relations with humans, and between humans through them. Here, the general framework has been likened to a constant, progressive and mutual mirroring process between the user and the device, benefiting the overall hybrid system by leveraging on the vulnerabilities of its components.

References

ACM. (2017). *Statement on algorithmic transparency and accountability*. Retrieved March 24, 2024, from https://www.acm.org/binaries/content/assets/public-policy/2017_usacm_statement_algorithms.pdf

Adadi, A., & Berrada, M. (2018). Peeking inside the black-box: A survey on explainable artificial intelligence (XAI). *IEEE Access, 6*, 52138–52160.

Aristotle Crisp, R. (Ed.). (2004). *Nicomachean ethics*. Cambridge University Press.

Armstrong, D. M. (1989). *Universals. An opinionated introduction*. Routledge.

Asilomar AI Principles. (2017). *Principles developed in conjunction with the 2017 Asilomar conference [Benevolent AI 2017]*. Retrieved March 24, 2024, from https://futureoflife.org/aiprinciples

Atomium European Institute. (2018). *AI4People's ethical framework for a good society: Opportunities, risks, principles and recommendations*. Retrieved March 24, 2024, from https://eismd.eu/featured/ai4peoples-ethical-framework-for-a-good-ai-society/

Attard-Frost, B., De los Ríos, A., & Walters, D. R. (2022). The ethics of AI business practices: A review of 47 AI ethics guidelines. *AI and Ethics, 3*(2), 1–18. https://doi.org/10.1007/s43681-022-00156-6

Attardi, G. (2023, June, 1–16). *Il Bello, il Brutto e il Cattivo dei LLM*. Mondo Digitale.
Balbi, A. (2021). *Inseguendo un raggio di luce. Alla scoperta della teoria della relatività*. Mondadori Libri S.p.A.
Beauchamp, T., & Childress, J. F. (2019). *Principles of biomedical ethics*. Oxford University Press.
Benanti, P. (2022). *Human in the loop. Decisioni umane e intelligenze artificiali*. Mondadori Education.
Bianca, M. (1995). Diritto Civile. *Vol.* 5–La Responsabilità. : Dott. A. Giuffré Editore S.p.A.
Bietti, E. (2021). From ethics washing to ethics bashing. A view on tech ethics from within moral philosophy. *Journal of social Computing, 2*(3), 266–283. https://doi.org/10.23919/JSC.2021.0031
Boddington, P. (2017). *Towards a code of ethics for artificial intelligence*. Springer.
Breiman, L. (2001). Statistical modeling: The two cultures. *Statistical Science, 16*(3), 199–231.
Buber, M. (1970). In W. Kaufmann (Ed.), *I and thou. A new translation with a prologue "I and you" and notes by*. Charles Scribner's Sons. (Original work published 1923).
Cachat-Rosset, G., & Klarsfeld, A. (2023). Diversity, equity, and inclusion in artificial intelligence: An evaluation of guidelines. *Applied Artificial Intelligence, 37*(1). https://doi.org/10.1080/08839514.2023.2176618
Chesterman, S. (2020). Artificial intelligence and the problem of autonomy. *Notre Dame Journal of Emerging Technologies, 1*.
Cian, G., & Trabucchi, A. (1992). *Commentario breve al Codice Civile. Complemento giurisprudenziale*. Casa Editrice Dott. Antonio Milani.
Coeckelbergh, M. (2013). *Human being @ risk: Enhancement, technology, and the evaluation of vulnerability transformations*. Springer.
Coeckelbergh, M. (2020). *AI ethics*. The MIT Press.
Correa, N. K., Galva, C., Santos, J. W., Del Pino, C., Pontes Pinto, E., Barbosa, C., Massman, D., Mambrini, R., Galvão, L., Terem, E., & de Oliveira, N. (2023). Worldwide AI ethics: A review of 200 guidelines and recommendations for AI governance. *Patterns, 4*, 100857. https://doi.org/10.1016/j.patter.2023.100857
Council of Europe. (2017). *Guidelines on the protection of individuals with regard to the processing of data in a world of Big Data*. T-PD(2017)01. Strasbourg. Retrieved March 24, 2024, from https://rm.coe.int/16806ebe7a
Crawford, K. (2021). *Atlas of AI: Power, politics, and the planetary costs of artificial intelligence*. Yale University Press.
Crawford, K., Dobbe, R., Dryer, T., Fried, G., Green, B., Kaziunas, E., Kak, A. et al. (2019). *AI now 2019 report*. Retrieved December 18, 2019, from https://ainowinstitute.org/AI_Now_2019_Report.pdf
Dadà, S. (2022). *Etica della vulnerabilità*. Morcelliana.
Descartes, R. (2008). In M. Moriarty (Ed.), *Meditations on first philosophy. With selections from the objections and replies*. Oxford University Press. (Original work published 1642).
Di Bella, S., & Schmaltz, T. M. (2017). *The problem of universals in early modern philosophy*. Oxford University Press.
Doshi-Velez, F., & Kim, B. (2017). Towards A rigorous science of interpretable machine learning. *arXiv:1702.08608v2 [stat.ML]*.
Dubber, M. D., Pasquale, F., & Das, S. (Eds.). (2020). *The Oxford handbook of ethics of AI*. Oxford University Press.
EGE. (2018). *Statement on artificial intelligence, robotics and 'autonomous' systems, published by the European Commission's European group on ethics in science and new technologies*. Retrieved March 24, 2024, from https://op.europa.eu/en/publication-detail/-/publication/dfebe62e-4ce9-11e8-be1d-01aa75ed71a1
European Commission. Directorate-General for Communications Networks, Content and Technology. (2019). *Ethics guidelines for trustworthy AI*. Publications Office. Retrieved March 24, 2024, from https://data.europa.eu/doi/10.2759/346720

European Union. (2012). Charter of Fundamental Rights of European Union. *Official Journal of the European Union. 2012/C 326/02*, C 326/391, Oct. 26th. Retrieved November 11, 2023, from https://eur-lex.europa.eu/legal-content/EN/TXT/?uri=celex%3A12012P%2FTXT

Fabris, A. (2012). *Etica delle nuove tecnologie*. Editrice La Scuola.

Fabris, A. (2016a). Philosophy, images and the mirror of machines. In Z. Paić & K. Purgar (Eds.), *Theorizing images* (pp. 111–120). Cambridge Scholars Publishing.

Fabris, A. (2016b). *RelAzione. Una filosofia performativa*. Editrice Morcelliana.

Fabris, A. (2018). *Ethics of information and communication technologies*. Springer.

FAT/ML. (2016). *Principles for accountable algorithms and a social impact statement for algorithms*. Retrieved March 24, 2024, from https://www.fatml.org/resources/principles-for-accountable-algorithms

Ferrarese, E. (2017). *The politics of vulnerability*. Routledge.

Ferraris, M. (2022, March 21–23). Perché i computer non possono morire. Cycle of seminars held at the *Istituto Italiano per gli Studi Filosofici*, Naples. Retrieved February 24, 2024, respectively from https://www.youtube.com/watch?v=K3LNnS5EPeg, https://www.youtube.com/watch?v=5MNVBDtw9kc, https://www.youtube.com/watch?v=Ah4qQVItQgM

Fjeld, J., Achten, N., Hilligoss, H., Nagy, A., & Srikumar, M. (2020). *Principled artificial intelligence: Mapping consensus in ethical and rights-based approaches to Principles for AI* (pp. 2020–2021). Berkman Klein Center Research Publication.

Floridi, L. (2014). *The fourth revolution. How the Infosphere is reshaping the world*. Oxford University Press.

Floridi, L. (2023). *The ethics of artificial intelligence. Principles, challenges and opportunities*. Oxford University Press.

Floridi, L., & Cowls, J. (2019). A unified framework of five Principles for AI in society. *Harvard Data Science Review, 1*(1). https://doi.org/10.1162/99608f92.8cd550d1

Floridi, L., Cowls, J., Beltrametti, M., Chatila, R., Chazerand, P., Dignum, V., Schafer, B., Luetge, C., Madelin, R., Pagallo, U., Rossi, F., Schafer, B., Valcke, P., & Vayena, E. (2018). AI4People–An ethical framework for a good AI society: Opportunities, risks, Principles and recommendations. *Minds and Machines, 28*, 689–707. https://doi.org/10.1007/s11023-018-9482-5

Frey, R. G., & Wellman, C. H. (2003). *A companion to applied ethics*. Wiley-Blackwell.

Galluzzo, G., & Loux, M. J. (2015). *The problem of universals in contemporary philosophy*. Cambridge University Press.

Genovesi, S., & Mönig, J. M. (2022). Acknowledging sustainability in the framework of ethical certification for AI. *Sustainability, 14*(7), 4157. https://doi.org/10.3390/su14074157

Gilligan, C. (1982). *In a different voice. Psychological theory and women's development*. Harvard University Press.

Gilson, E. (2014). *The ethics of vulnerability. A feminist analysis of social life and practice*. Routledge.

Giovanola, B., & Tiribelli, S. (2022). Weapons of moral construction? On the value of fairness in algorithmic decision-making. *Ethics and Information Technology., 24*, 3.

Goffi, E. R. (2019). *De l'éthique à la cosm-éthique: ce que l'éthique n'est pas*. Institut Sapiens.

Goffi, E. R., Colin, L., & Belouali, S. (2021). Ethical assessment of AI cannot ignore cultural pluralism: A call for broader perspective on AI ethics. *Arribat–International Journal of Human Rights, 1*, 151–175.

Guidotti, R., Monreale, A., Ruggieri, S., Turini, F., Pedreschi, D., Giannotti, F. (2018). *A survey of methods for explaining black box models*. https://doi.org/10.48550/arXiv.1802.01933.

Gwagwa, A., Kraemer-Mbula, E., Rizk, N., Rutenberg, I., & De Beer, J. (2020). Artificial intelligence (AI) deployments in Africa: Benefits, challenges and policy dimensions. *The African Journal of Information and Communication, 26*, 1–28. https://doi.org/10.23962/10539/30361

Gwagwa, A., Kachidza, P., Siminyu, K., & Smith, M. (2021). *Responsible artificial intelligence in sub-Saharan Africa: Landscape and general state of play*. AI4D Africa.

Hagendorff, T. (2020). The ethics of AI ethics: An evaluation of guidelines. *Minds and Machines, 3*, 99–120. https://doi.org/10.1007/s11023-020-09517-8

Hagendorff, T. (2022). Blind spots in AI ethics. *AI and Ethics, 2*, 851–867. https://doi.org/10.1007/s43681-021-00122-8

Helberger, N., Sax, M., Strycharz, J., et al. (2022). Choice architectures in the digital economy: Towards a new understanding of digital vulnerability. *Journal of Consumer Policy, 45*, 175–200. https://doi.org/10.1007/s10603-021-09500-5

Hickock, M. (2021). Lessons learned from AI ethics principles for future actions. *AI and Ethics, 1*, 41–47. https://doi.org/10.1016/j.patter.2023.10085

Hine, E., & Floridi, L. (2022). Artificial intelligence with American values and Chinese characteristics: A comparative analysis of American and Chinese governmental AI policies. *AI & Society*. Retrieved September 2, 2024 from, *39*, 257–278. https://doi.org/10.1007/s00146-022-01499-8

Holzinger, A., Langs, G., Denk, H., Zatloukal, K., & Mueller, H. (2019). Causability and explainability of artificial intelligence in medicine. *WIREs Data Mining and Knowledge. Discovery, 9*., Article e1312, e1312. https://doi.org/10.1002/widm.1312

House of Lords Artificial Intelligence Committee. (2018). *AI in the UK (AIUK). Ready, willing and able.* Retrieved March 24, 2024, from https://publications.parliament.uk/pa/ld201719/ldselect/ldai/100/10002.htm

IBM. (2017). *Transparency and trust in the cognitive era.* Retrieved March 24, 2024, from https://www.ibm.com/blogs/southeast-europe/transparency-trust-cognitive-era/

IEEE. (2017). *Ethically aligned design: a vision for prioritizing human well-being with autonomous and intelligent systems, Version 2.* Retrieved March 24, 2024, from https://standards.ieee.org/wp-content/uploads/import/documents/other/ead_v2.pdf

IEEE. (2019). *IEEE global initiative on ethics of autonomous and intelligent systems. Ethically aligned design: A vision for prioritizing human well-being with autonomous and intelligent systems*. First Edition. Retrieved March 24, 2024, from https://standards.ieee.org/industry-connections/ec/ead1e-infographic/

Jobin, A., Ienca, M., & Vayena, E. (2019). Artificial intelligence: The global landscape of ethics guidelines. *Nature Machine Intelligence, 1*, 389–399.

Kahneman, D. (2011). *Thinking, fast and slow*. Farrar, Straus and Giroux.

Kiemde, S. M. A., & Kora, A. D. (2022). Towards an ethics of AI in Africa: Rule of education. *AI and Ethics, 2*, 35–40. https://doi.org/10.1007/s43681-021-00106-8

Kizza, J. M. (2009). Computer network vulnerabilities. In J. M. Kizza (Ed.), *A guide to computer network security. The computer communications and networks*. Springer.

Laarson, S. (2020). On the governance of artificial intelligence through ethics guidelines. *Asian Journal of Law and Society, 00*, 1–15. https://doi.org/10.1017/als.2020.19

Larsson, S., & Heintz, F. (2020). Transparency in artificial intelligence. *Internet Policy Review, 9*(2). https://doi.org/10.14763/2020.2.1469

Lipton, Z. (2016). The mythos of model interpretability. ICML workshop on human interpretability in machine learning (WHI 2016). .

Liu, H. W., & Lin, C. F. (2020). Artificial intelligence and global trade governance: A pluralist agenda. *Harvard International Law Journal, 61*(2). Retrieved March 24, 2024, from SSRN: https://ssrn.com/abstract=3675505

Loi, M., Heitz, C., & Christen, M. (2020). A comparative assessment and synthesis of twenty ethics codes on AI and big data. In *2020 7th Swiss conference on data science (SDS)*, 41–46. https://doi.org/10.1109/SDS49233.2020.00015.

Luengo-Oroz, M. (2019). Solidarity should be a core ethical principle of AI. *Nature* Machine Intelligence, 494. Retrieved March 24, 2024, from https://www.nature.com/articles/s42256-019-0115-3

Madras, D., Pitassi, T., & Zemel, R. (2018). Predict responsibly: Improving fairness and accuracy by learning to defer. *arXiv:1711.06664v3 [stat.ML].*

Malgieri, G., & Jędrzej, N. (2020). Vulnerable data subjects. *Computer Law and Security Review, 37*, 105415.

Mancilla-Caceres, J. F., & Estrada-Villalta, S. (2020). The ethical considerations of AI in Latin America. *DISO, 1*(16), 1–6. https://doi.org/10.1007/s44206-022-00018-y

Marda, V., & Acharya, B. (2014). *Identifying aspects of privacy in Islamic law*. The Centre for Internet and Society. Retrieved March 24, 2024, from https://cis-india.org/internet-governance/blog/identifying-aspects-of-privacy-in-islamic-law

Metzinger, T. (2019, November 2). *Ethics washing made in Europe*. Der Tagesspeigel. Retrieved March 24, 2024, from https://www.tagesspiegel.de/politik/eu-guidelines-ethics-washing-made-in-europe/24195496.html

Mhlambi, S. (2020). From rationality to relationality: Ubuntu as an ethical and human rights framework for artificial intelligence governance. *Carr Center for Human Rights Policy Discussion Paper Series, 9*, 1–27.

Microsoft. (2018). *Responsible bots: 10 guidelines for developers of conversational AI*. Retrieved September 6, 2024, from https://www.microsoft.com/en-us/research/publication/responsible-bots/

Miller, T. (2023). Explainable AI is dead, long live explainable AI! *arXiv:2302.12389v3 [cs.AI]*.

Misselhorn, C. (2022). Artificial moral agents. Conceptual issues and ethical controversies. In S. Voeneky, P. Kellmeyer, O. Mueller, & W. Burgard (Eds.), *The Cambridge handbook of responsible artificial intelligence*. Cambridge University Press.

Mittelstadt, B. (2019). Principles alone cannot guarantee ethical AI. *Nature Machine Intelligence, 1*, 501–507. https://doi.org/10.1038/s42256-019-0114-4

Mohamed, S., Ping, M. P., & Isaac, W. (2020). Decolonial AI: Decolonial theory as sociotechnical foresight in artificial intelligence. *Philosophy & Technology, 33*, 659–684.

Moor, J. H. (2006). The nature, difficulty and importance of machine ethics. *IEEE Intelligent Systems, 21*(2006), 18–21.

Natale, S. (2021). *Deceitful media. Artificial intelligence and social life after the Turing test*. Oxford University Press.

Novelli, C., Taddeo, M., & Floridi, L. (2023). Accountability in artificial intelligence: What it is and how it works. *AI & Society, 39*, 1871–1882. https://doi.org/10.1007/s00146-023-01635-y

Ntoutsi, E., Fafalios, P., Gadiraju, U., Iosifidis, V., Nejdl, W., Vidal, M. E., Ruggieri, S., Turini, F., Papadopoulos, S., Krasanakis, E., Kompatsiaris, I., Kinder-Kurlanda, K., Wagner, C., Karimi, F., Fernandez, M., Alani, H., Berendt, B., Kruegel, T., Heinze, C., Broelemann, K., Kasneci, G., Tiropanis, T., & Staab, S. (2020). Bias in data-driven artificial intelligence systems–An introductory survey. *WIREs Data Mining Knowl Discov, 10*, Article e1356. https://doi.org/10.1002/widm.1356

O'Neil, C. (2016). *Weapons of math destruction: How big data increases inequality and threatens democracy*. Broadway Books.

OECD. (2019). *Recommendation of the council on artificial intelligenc*e. Retrieved March 24, 2024, from https://legalinstruments.oecd.org/en/instruments/OECD-LEGAL-0449

Pap, A. (1943). On the meaning of universality. *The Journal of Philosophy, 40*(19), 505–514.

Pipkin, D. (2000). *Information security: Protecting the global Enterprise*. Prentice Hall PTR.

Preece, A. (2018). Asking "why" in AI: Explainability of intelligent systems–Perspectives and challenges. *Intelligent Systems in Accounting, Finance and Management, 25*, 63–72. https://doi.org/10.1002/isaf.1422

Punzi, C., Setzu, M., Pellungrini, R., Giannotti, F., & Pedreschi, D. (2023). *Towards synergistic human-AI collaboration in hybrid decision-making systems*. Retrieved March 24, 2024, from https://scholar.google.com/citations?view_op=view_citation&hl=en&user=3UL0kVUAAAAJ&sortby=pubdate&citation_for_view=3UL0kVUAAAAJ:hMod-77fHWUC

Razmi, F., Lou, J., Hong, Y., & Xiong, L. (2023, September 18–22). Interpretation attacks and defenses on predictive models using electronic health records. In D. Koutra, C. Plant, M. Gomez Rodriguez, E. Baralis, & F. Bonchi (Eds.), *Machine learning and knowledge discovery in databases. Research track. Proceedings of the European conference ECML PKDD 2023*.

RenAIssance Foundation. (2020). *The Rome call for AI ethics*. Retrieved May 14, 2024, from https://www.romecall.org/the-call/

Robertson, L. J., Abbas, R., Alici, G., Munoz, A., & Michael, K. (2019). Engineering-based design methodology for embedding ethics in autonomous robots. *Proceedings of the IEEE, 107*(3), 582–599.

Roche, C., Lewis D., & Wall, P. J. (2021). *Artificial intelligence ethics: An inclusive global discourse?* Retrieved March 24, 2024, from https://arxiv.org/pdf/2108.09959.pdf

Rosenblueth, A., Wiener, N., & Bigelow, J. (1943, January). Behavior, purpose and teleology. *Philosophy of Science, 10*(1), 18–24.

Rudschies, C., Schneider, I., & Simon, J. (2020). Value pluralism in the AI ethics debate—Different actors, different priorities. *AI, Ethics and Society, 29*, 1–15. https://doi.org/10.29173/irie419

Ryan, M., & Stahl, B. C. (2020). Artificial intelligence ethics guidelines for developers and users: Clarifying their content and normative implications. *Journal of Information Communication and Ethics in Society, 19*(1). https://doi.org/10.1108/JICES-12-2019-0138

Santoni de Sio, F., & Mecacci, G. (2021). Four responsibility gaps with artificial intelligence: Why they matter and how to address them. *Philosophy and Technology, 34*, 1057–1084. https://doi.org/10.1007/s13347-021-00450-x

Schiaffonati, V., & Tamburrini, G. (2021). Armi autonome e controllo umano significativo. In F. Fossa, V. Schiaffonati, & G. Tamburrini (Eds.), *Automi e persone. Introduzione all'etica dell'intelligenza artificiale e della robotica* (pp. 133–152). Carocci editore S.p.A.

Schiff, D., Borenstein, J., Biddle, J., & Laas, K. (2021). AI ethics in the public, private, and NGO sectors: A review of a global document collection. *IEEE Transactions on Technology and Society, 1*, 31–42.

Schiff, D. S., Laas, K., Biddle, J. B., & Borenstein, J. (2022). Global AI ethics documents: What they reveal about motivations, practices, and policies. In Laas et al. (Eds.), *Codes of ethics and ethical guidelines: Emerging technologies, changing field* (pp. 121–143). Springer.

Schwetz, B. A., & Toulmin, S. (2004, May 14). *Oral history of the Belmont report and the National Commission for the protection of human subjects of biomedical and Behavioural research.* Retrieved March 24, 2024, from https://www.hhs.gov/ohrp/education-and-outreach/luminaries-lecture-series/belmont-report-25th-anniversary-interview-stoulmin/index.html

Searle, J. (1980). Minds, brains and programs. *The Behavioral and Brain Sciences, 3*, 417–457.

Seger, E. (2022). In defence of principlism in AI ethics and governance. *Philosophy & Technology, 35*. https://doi.org/10.1007/s13347-022-00538-y

Skitka, L. S., Mosier, K. L., & Burdick, M. (1999). Does automation bias decision making? *International Journal of Human-Computer Studies, 51*, 991–1006.

Stolk, A., Verhagen, L., & Toni, I. (2016). Conceptual alignment: How brains achieve mutual understanding. *Trends in Cognitive Sciences, 20*(3), 180–191. https://doi.org/10.1016/j.tics.2015.11.007

Szegedy, C., Zaremba, W., Sutskever, I., Bruna, J., Erhan, D., Goodfellow, I., Fergus, R. (2013). Intriguing properties of neural networks. *arXiv*:1312.6199v4 [cs.CV].

Telefonica. (2018). *AI principles of Telefonica*. Retrieved March 24, 2024, from https://www.telefonica.com/en/wp-content/uploads/sites/7/2021/11/principios-ai-eng-2018.pdf

Ten Have, H. (2016). *Vulnerability. Challenging bioethics*. Routledge.

The Partnership on AI. (2016). *Tenets*. Retrieved March 24, 2024, from https://partnershiponai.org/about/#tenets

Turilli, M., & Floridi, L. (2009). The ethics of information transparency. Ethics and Information Technology, 11, 105–112. Retrieved August 31, 2024, from https://doi.org/10.1007/s10676-009-9187-9.

Turing, A. (1950). Computing machinery and intelligence. *Mind, 49*(1950), 433–460.

UK Government. Department for digital, culture, media and sport. (2018). *Data ethics framework*. Retrieved March 24, 2024, from https://assets.publishing.service.gov.uk/media/5f74a4958fa8f5188dad0e99/Data_Ethics_Framework_2020.pdf. Available through *A guide to good practice for digital and data-driven health technologies*. Retrieved March 24, 2024, from https://www.gov.uk/

government/publications/code-of-conduct-for-data-driven-health-and-care-technology/initial-code-of-conduct-for-data-driven-health-and-care-technology

University of Montreal. (2017). *Montreal declaration for a responsible development of artificial intelligence*. Retrieved March 24, 2024, from www.montrealdeclaration-responsibleai.com/the-declaration

Van Norren, D. E. (2023). The ethics of artificial intelligence, UNESCO and the African Ubuntu perspective. *Journal of Information, Communication and Ethics in Society, 21*(1), 112–128. https://doi.org/10.1108/JICES-04-2022-0037

von Eschenbach, W. J. (2021). Transparency and the black box problem: Why we do not trust AI. *Philosophy and Technology, 34*, 1607–1622. https://doi.org/10.1007/s13347-021-00477-0

Wagner, B. (2018). Ethics as an escape from regulation. From "ethics-washing" to ethics-shopping? In E. Bayamlioglu, I. Baraliuc, L. A. W. Janssens, & M. Hildebrandt (Eds.), *Being profiled: Cogitas ergo sum: Cogitas ergo sum: 10 years of profiling the European citizen* (pp. 84–89). Amsterdam University Press. https://doi.org/10.1515/9789048550180-016

Wallach, W., & Allen, C. (2009). *Moral machines: Teaching robots right from wrong*. Oxford University Press.

Webb, T., Holyoak, K. J., & Lu, H. (2023). Emergent analogical reasoning in large language models. *Nature Human Behaviour, 7*, 1526–1541. https://doi.org/10.1038/s41562-023-01659-w

Wei, J., Tay, Y., Bommasani, R., Raffel, C., Zoph, B., Borgeaud, S., Yogatama, D., Bosma, M., Zhou, D., Metzler, D., Chi, E. H., Hashimoto, T., Vinyals, O., Liang, P., Dean, J., & Fedus, W. (2022 August). Emergent abilities of large language models. *Transactions on Machine Learning Research*. arXiv:2206.07682v2 [cs.CL].

Wiener, N. (1964). God and golem. In *Inc. A comment on certain points where cybernetics impinges on religion*. The M.I.T. Press.

Wong, P. H. (2020). Cultural differences as excuses? Human rights and cultural values in global ethics and governance of AI. *Philosophy and Technology, 33*, 705–715. https://doi.org/10.1007/s13347-020-00413-8

Wright, A. P., Wang, Z. J., Park, H., Guo, G., Sperrle, F., Ell-Assady, M. Endert, A., Keim, D., & Chau, D. H. (2020). A comparative analysis of industry human-AI interaction guidelines, *arXiv*:2010.11761. https://doi.org/10.48550/arXiv.2010.11761

Zeng, Y., Lu, E., & Huangfu, C. (2019). Linking artificial intelligence Principles, proceedings of the AAAI workshop on artificial intelligence safety, arXiv:1812.04814 [cs.AI]. https://doi.org/10.48550/arXiv.1812.04814

Zhang, X. Y., Xie, G. S., Li, X., Mei, T., & Liu, C. L. (2023, February). A survey on learning to reject. *Proceedings of the IEEE, 111*(2). https://doi.org/10.1109/JPROC.2023.3238024

Ethics of Digital Technologies: Where We Are, Where We Are Heading Towards

How Humans Can Flourish in Digital Environments

Giampaolo Ghilardi and Paola Iadaresta

Abstract The aim of the chapter is to explore the anthropological and moral implications connected with the new landscape digital technologies are bringing about. The digital era is changing our idea of ourselves and the world: we will explore the change that the notion (and the reality) of the environment has undergone, passing form a naturalistic concept to a synthetic one. The fuel of digital environment is data: therefore the "digital life" can be understood in terms of the rational model of *Dataism*—with its logic of correlation as opposed to the causal explanation—and its theoretical instrument, the macroscope. Since computation is not neutral action but instead embedded with choices, decisions and evaluation, we will try to explore its ethical dimension, that goes under the name of *Algorethics*. As a sort of synthesis and case study of ideas reviewed we shall see the concept of On-life, that keeps together the digital dimension with the day-by-day life in a ubiquitous and perennial present. The ethical question we will try to address is how it is possible for the human being to flourish in such environment: specular to the digital environment there is a human reality to consider carefully, a nature which is inherently analogical rather than digital, where time and space cannot be compressed to the point to disappear in a continuum flow of presence.

Keywords Digital environment · Dataism · Algorethics · Onlife · Flourishing · Human ecology

In order to correctly attribute the work, the Authors specify that they both contributed equally to the architecture and to the overall composition of the chapter, as well as to its revision and conclusions, whereas in detail, Giampaolo Ghilardi dealt mainly with the paragraphs (1,3,4,5,6) while Paola Iadaresta with paragraphs (2,7,8).

G. Ghilardi · P. Iadaresta (✉)
Università Campus Bio-Medico di Roma, Rome, Italy
e-mail: g.ghilardi@unicampus.it; p.iadaresta@unicampus.it

A. Fabris, S. Belardinelli (eds.), *Digital Environments and Human Relations*, Human Perspectives in Health Sciences and Technology 150,
https://doi.org/10.1007/978-3-031-76961-0_3

1 Introduction

The aim of this chapter is to explore the anthropological and moral implications connected with new landscape digital technologies are bringing about. Differently from other kinds of technologies, the new digital era is indeed changing not just the environment we live in, which by itself is worth noting and analyzing, but moreover it is changing our behaviors, our perceptions of time and space, in short, our idea of ourselves and the world.

To achieve this task, we will closely explore the change that the notion (and the reality) of environment has undergone, passing from a naturalistic concept to the actual synthetic one. We will reason about the physical constitution of the digital environment, together with its virtual nature. The analysis will lead us to consider this new phenomenon that we could call the primacy of potency on actuality, the opposite of the Aristotelian philosophy.

The reasoning will step forward then on the curious idea of Dataism. Data may be considered somehow the real fuel of digital environment, as a matter of fact the very unity of measure today of the new scientific environment is Data. Dataism therefore will be studied as the new ideology, or framework if the previous word should sound too strong, behind the "digital life". It will be important to understand what kind of rational model stays at the ground of this new scientific style. The hot topic on this paragraph will be the study of the switch that is taking place between the causal explanation and the correlation.

Once we have assessed the nature of Dataism, we should be able to understand the other important tool for this new environment: the macroscope, that is the theoretical instrument enabling the user to capture layers of reality, through Data correlation, otherwise not available. This powerful instrument needs to be carefully understood so to be in turn appropriately adopted. Without a deep knowledge of its nature and its purpose, we risk mistaking its instrumental essence, confusing the means with the ends.

To this point we will face the new coined concept of Algorethics, the ethics of algorithms, or better said the ethics of computations. Since algorithms are nothing but a set of instruction, we will try to understand how is that a set of rules can become of ethical concern. We could discover that computation, after all, is not a so neutral action after all, embedded with choices, decisions, and evaluation. Once it has been granted the proper level of understanding of this sphere, it will flow smoothly the idea that computing requires much more than just arithmetical skills: it requires the capacity to see the world the calculus opens up, to see if it fits the reality numbers would like to describe.

At this point we shall see the on-life concept as a sort of synthesis and case study of ideas reviewed. In fact, this made-up word keeps together the digital, online, dimension with the day-by-day life, which is the focus of the chapter. The challenge is therefore how to apply this blurring distinction to a new world that seems to have undergone through important changes. We'll try to see the ontological side of the question, how being on life entails a sort of return to Parmenides: paradoxically in a world full of virtual reality, the very idea of potency is somehow put aside in favor to a ubiquitous and perennial present.

In this scenario Sects. 7 and 8 will tackle the short and long-term consequences to be drawn from these premises. The ethical question we will try to address is how it is possible in such environment for the human being to thrive and even rejoice, in a word: to flourish. Is it still an option to follow a path that eventually will lead human beings to be happy? To fulfill their potentials? To answer these questions, we'll have to look to what has been named as human ecology. Specular to the digital environment, there is the human reality to consider carefully, a nature which is inherently analogical rather than digital, where time a space cannot be compressed to the point to disappear in a continuum flow of presence. To live onlife, somehow seems to disrupt the material constitution of man, to disregard his natural constrains, to overlook his nature.

2 Environment vs Digital Environment

Descending from the Middle French preposition *environ* "around," environment is "that which surrounds." The Oxford Dictionary defines it as «the physical surroundings or conditions in which a person or other organism lives, develops, etc., or in which a thing exists; the external conditions in general affecting the life, existence, or properties of an organism or object». The prefix -en refers to a spatial dimension. Therefore, the environment is the physical dimension in which an organism is situated—*the dimension that materially surrounds it*. The environment plays a crucial role in shaping the perception and understanding of the world: through the senses, an embodied organism perceives the surrounding space, moves within it, and interacts with the objects and other individuals populating it. Physical space thus provides a tangible context in which to experience, relate, and engage in meaningful actions. Bodily experience occurs not only in space but also in time: through its perception, an organism feels the flow of events, organizes its experiences into a sequence of distinct moments, and consciously engages the present. The embodied experience of an organism occurs at the intersection of the spatiality of its body and its environment and the flow of time—with all the changes it entails—to which that body and environment are subject.

Given these assumptions, is it plausible to speak of "digital environments" (Boczkowski & Mitchelstein, 2021) if the digital realm lacks physical reality—hence space-time, but we, on the other hand, still experience it? The digital environment is a numerical, mathematical one, which, although not "material" in the traditional sense of the term, still allows us, as embodied subjects, to have experiences within it. What kind of experience is, therefore, that of an environment that materially does not exist?

The digital environment certainly has a physical foundation: this consists of servers, networks, devices, and software—which together provide the infrastructure necessary for digital existence and interaction. But the flow of data and information that make up online platforms, social media, and other forms of digital technology, while in some way "existing," cannot be "touched"– they remain *intangible*. They lack spatial-temporal reality because the number—the digit—is an abstraction that has neither space nor time. The binary representations 0 and 1 used to encode, process, transmit, and store data have no material reality and create a 'immaterial' mathematical environment with which we can nevertheless interact. The traditional notion of environment, rooted in space and time, is insufficient to characterize the digital reality. This does not mean that space and time are foreign to the digital, but that in the digital they exist in a different form.

Ben Agger coined an interesting expression to define digital time: *iTime*. iTime represents how digital technologies have radically transformed the way people manage time and activities. With smartphones, people are constantly connected to work, communication, and entertainment, both during and outside traditional working hours. iTime frees the subject from the constraints of space and time: in the digital environment, instead of being tied to physical space, the subject is everywhere, at any time, always connected, and above all always *available*. Thanks to mobile technologies, virtual space becomes an extension of physical space and allows people to work, communicate, and entertain wherever there is an internet connection. If, on the one hand, iTime is an eternal present that requires constant availability, iSpace (an expression that I infer from Agger, although he does not use it) is that space that makes it possible to respond, always and everywhere, to this continuous demand for availability. Real spaces begin to overlap: you can respond to a work email when you are in the park with your family—and not only and exclusively when you are in your office. The traditional concept of place as a reference point for human activities thus becomes increasingly irrelevant.

The digital environment therefore requires continuous participation: «iTime feels infinite, never to be bracketed by beginnings and endings» (Agger, 2011). The online world and the offline world, in digital time, overlap. In 2012, under the supervision of Luciano Floridi, a group of scholars drafted a document entitled "The Onlife Manifesto" in which this word, "onlife," indicates the complex interaction between humanity and technology, between human life and the digital environment—which in onlife form a single integrated reality: «We decided to adopt the neologism "onlife" that I had coined in the past in order to refer to the new experience of a hyperconnected reality within which it is no longer sensible to ask whether one may be online or offline» (Floridi, 2015). By infinitely dilating space and time, the digital offers possibilities that were unthinkable until a few decades ago. But the ubiquity and always possible availability in onlife do not imply nor equate to true presence—because presence is bodily and spatio-temporal. The word "present" derives from the Latin *prae*, before, and *sum*, *I am*: I am here, materially present in front of someone else. The prefix pre- has a preventing cause function: being here, in front of you, I cannot be elsewhere. Presence implies that to be at point x I cannot also be at point y or z: the principle of non-contradiction would somehow be violated. In the availability of iTime, however, the subject is not truly present neither here nor there but potentially available to be both here and there. It is at the same time *everywhere and nowhere*. Availability, in iTime, does not require a change of place or a division of time—because everything is fused into an eternal present and identical space. A present that never implies—and this is a contradiction in terms if we consider the etymology of the word 'present'—true presence because true presence requires being here, in this moment, without the possibility, by exclusion, of being elsewhere. *The digital environment does not require full presence but continuous availability*. A question worth trying to answer is the following: our embodied nature implies rootedness in space-time. The digital possesses altered space-time characteristics, outside our traditional way of experiencing. Is it therefore possible to be present—in the material sense of the term—in an environment, the digital one, that is neither material nor embodied?

In order to answer this question though, let's take a look at another important feature of the digital environment, that is its rootedness in the world of data: the Dataism.

3 Dataism

> If you asked me to describe the rising philosophy of the day, I'd say it is data-ism. We now have the ability to gather huge amounts of data. This ability seems to carry with it certain cultural assumptions —that everything that can be measured should be measured; that data is a transparent and reliable lens that allows us to filter out emotionalism and ideology; that data will help us do remarkable things —like foretell the future (Brooks, 2013).

In these few lines, written by David Brooks on the New York Times in 2013, we find the first occurrence of the neologism *Dataism*, a mixed-up word where we can see the merge of two different and apparently very distant domains: data and art. Dadaism, in fact, is a well-known contemporary art movement, developed in reaction to World War I, it mainly consisted of artists, refugees based in Zurich, who rejected logic, reason, and aestheticism of modern capitalist society, instead expressing nonsense, irrationality, and anti-bourgeois protest in their works. So, on one side we have data, the quintessential product of rational work, strictly logical connections, and algorithms; on the other side we have an avant-garde art, devoted to object rationalism, and yet the idea of data-ism makes sense, since somehow brings together the perfectly deterministic working way of data and the not so rational, in fact quite arbitrary, colonization of the use of these big-data in our daily routine.

So, Dataism, broadly speaking (Bertolaso et al., 2022; Fuchs, 2022), becomes the leading ideology of the time where data appears to be the only relevant element of the decision process, in an all too predictable framework of choices. But where is the Dadaistic aspect of this "tyranny" of data? In a nutshell, we might say that it dwells in the pretension to use it in every possible domain of human activity. It's not the data itself the issue, of course, but it's over-usage, it's the way we human approach the data that is non so logic anymore. As it would be not logic to use a telescope to see someone face.

As it often happens, the ethical issue at stake when it comes to technology relies on at least two aspects: the object itself, of course, and above all the use we make of it. The object is never neutral, as sometimes we hear to say, since it had been brought to the light by different agents with different agendas, that can, and therefore should, be taken into account when we want to morally reason about it. The use we make of the artifact then can be appropriate or not, depending both on the nature of the object, and on the nature of the user, who undergoes through specific changes by using this kind of devices.

A curious example given by Prof. Paolo Benanti about some hypothetical misuse of data is the following (Benanti, 2017). Let's say we need to choose between two fiancés Annah and Mary, we like them both each of them for different reasons, so we cannot decide and ask "Google" to help us in this choice. AI could answer us that it knows us since the day we were born, it knows every email exchange we had with them, every biometrical reaction, every conversation, every kind of social media interaction we might have had. And the same applies to them, so that this hypothetical AI know us and them better than we could possibly imagine, *therefore* it is most entitled to give us the right advice on the matter, even with success rate probability.

This funny portrait is not so unconceivable after all,[1] it's an everyday experience today to find ourselves to be read and anticipated by our mobile: whenever we move at a certain time and toward a certain direction, out of nowhere, even if we don't ask anything, the internal mobile AI diligently opens for us the GPS, showing the desired path without any query, and more often than not it really is our intended destination, grasped by behavioral path gathered the days before and crossed with temporal data as well.

It is no wonder then that this new cult for data is becoming a sort of scientific faith. No more than 16 years ago, Chris Anderson, the Wired director said:

> Scientists have always relied on hypotheses and expressions. [...] Faced with the availability of enormous quantities of data, this approach – hypothesis, theoretical model and test – becomes obsolete. [...] There is now a better way. Petabytes allow us to say: "correlation is enough". We can stop looking for theoretical models. We can analyze the data without any assumptions about what it might show. We can send the numbers into the largest cluster of computers the world has ever seen and let statistical algorithms find patterns where science cannot. [...] Learning to use a computer of this scale can be challenging. But the opportunity is great: the new availability of an enormous amount of data, combined with the statistical tools to process it, offers a completely new way to understand the world. Correlation supplants causation and science can advance even without coherent theoretical models, unified theories or some kind of mechanistic explanation (Anderson, 2008).[2]

This idea that correlation is enough, and that we can celebrate therefore the obituary of explanation or interpretation is naïve and wrong. Naïve, because the combination of data and statistical tools aforementioned above is of course nothing but an interpretation, Statistics is not a neutral theory, there are different kinds of statistics, and to adopt one instead of another is a methodological choice, not data-driven, even if

[1] To see how these scenarios are getting more and more common nowadays, see (Crippa & Girgenti, 2024).

[2] Translation of mine.

it aims to make sense of data. Wrong, because correlation may be the beginning of scientific reasoning, it gives to the scientist the matter to delve in, but until a certain level of causality is not reached, we don't have the necessary epistemic stability needed to frame a scientific sentence. It is not enough to say that A occurs when B happens, we should be able to say why these two different phenomena are correlated one another. Without this stronger form of relationship, we remain on the domain of the observation, we don't gain the level of the explanation, which is the specific hallmark of scientific reasoning.

4 The Macroscope

The availability of a large amount of data *per se* would not be enough to produce what we named as Dataism. In order to have this sort of imperialism of data, *big* data, in our life we also need to have the proper instrument to make sense of this ever-growing quantity of numbers. This instrument is called macroscope. Differently from the other scopes (it comes from the Greek verb σκοπέω, *skopew*, that means "to observe"), e.g. telescope and microscope, which were variations of the specific ability to craft curved lens to enhance our vision, this one is not made of lens, it does not depend on our ability to mold the glass, or more broadly the matter, the macroscope is more a conceptual tool than a material one. It is precisely the way to look at big scale event, otherwise not detectable if it were not for the data analysis. In the words of Joël de Rosnay, one of the first to coin and use the term:

> Let us use the macroscope to direct a new look at nature, society, and man and to try to identify new rules of education and action. In its field of vision organizations, events, and evolutions are illuminated by a totally different light. The macroscope filters details and amplifies that which links thing together. It is not used to make things larger or smaller but to observe what is at once too great, too slow, and too complex for our eyes (human society, for example, is a gigantic organism that is totally invisible to us) (De Rosnay, 1979).

So, in simpler words, the macroscope is the ability that we develop through computers, algorithms, and analytical system to look into the data and seeing patterns otherwise non observable. This ability, as we can see, does not depend on the physical nature of the instrument we are using, it rather relies on its formal structure. It's more a sort of calculus that allows us to see what we could not see otherwise. The macroscope therefore allows us a sort of conceptual vision rather than an empirical one, and yet it depends completely on the data it is built on, to accomplish its task. So, we find ourselves in a kind of paradoxical situation, where the instrument we are using is a formal way of linking numbers together, but at the same time these *numbers* are not abstract entities, they should be the formalization of the data extracted from the digital sources. From an epistemological point of view, it is not a simple task to distinguish whether, and to what extent, we are talking about something real, existing out there, or about something abstract, with little if any connection to the reality. It is difficult to assess the range of the empiricism implied in this new kind of technology.

At a first sight, in fact, it would appear that the very nature of this instrument, being substantially a way of calculus, should be the one typical of formal objects, rationalistic, mathematical; but then we are told that data, the numbers the macroscope works with, are not just numbers, they are *real* representations of the world we live in, hence the ambiguity of the tool: part mathematical, part empirical; of course we need to understand the range of this empirical part, since the relation between the numbers and the reality they are expressing is not something that could possibly be considered stable. At its best this relation has a probabilistic nature, and it cannot be thought as a mirror of the phenomenon it is *portraying*.

By the way, with all these caveats in mind, we should recognize the value of such an instrument, that is hybrid in its composition, and allows us to see the deeper layers of our world, making it a more comprehensible environment. This point might need a little bit more of clarification. According to Scheler's Anthropology (Scheler, 2009) the main difference between the world and the environment is that the first one is an open category, while the second one is a closed space, a field plenty of affordances, rich of possibilities, and yet somehow something with a given and specific number of possible interactions. The world cannot be measured or counted, the environment can and should be. The environment can be thought as a downgraded world, a world that has lost its openness. Just humans have a world size gaze, while animals live in specific environment.

So, macroscope on the one hand helps us to downsize world into a more handling tool, it helps us to domesticate frightening or dangerous places; on the other hand, the achievement of this task deprives the world from its openness, losing therefore its specific human feature, its unpredictability, its source of novelty. Every "scope" somehow aims to this goal, macroscope reaches this purpose not just by the means of its scoping strength, but rather because of its digital nature. The algorithms the macroscope is made of allow us to anticipate future scenarios and prepare ourselves to engage with them. So that this technology, as in several case it is possible to observe, comes with a price; it can be a very powerful and useful tool, but it can also be a dehumanizing instrument. These two sides of the coin are embedded in the "instrumental" nature of the tool itself.

We would not say that the morality is all on the side of the final user, the *finis operantis* (the user goal) in the traditional words: there is also an important component of the tool *in se* which needs to be accounted for the morality to be considered. On this side it is possible to say that this tool provides us with a cognitive improvement that can be considered something good, *per se*; what then will be done of this cognitive improvement, that is up to the user to decide, and there lies the ethical part of our reasoning. Of course, in order to properly master and therefore correctly use these kinds of tools, we need to know how they work, but here we find a huge problem: even the developers of such tools more often than not do not know how these algorithms really work. When deep learning is involved, it is almost impossible to trace down all the operations that lead the machine to take a specific path rather than another one. This kind of opacity is precisely what prevent the user to be completely confident with the outcome of these tools. This is also the reason according to which it is not possible to rely completely and exclusively on these instruments. As we said, it is a problem that relies upon the nature of the tool, on its working mechanism, it is not an issue concerning its teleology.

5 Algorethics

This last order of considerations leads us to consider what has been defined as *Algorethics*: the field of ethics that focuses on the development and deployment of algorithms. It aims to ensure that algorithms are transparent, accountable, and fair. It also strives to involve and empower diverse stakeholders in the algorithmic decision-making process. The term has been introduced in 2018, indicating the necessity for a dedicated study concerning the evaluation of an ethics applied to technology, to Algorithms and to Artificial Intelligence (AI) (Mantini, 2022).

This new developing ethical field is about the "morality" of algorithms (Lindberg & Roine, 2021). Now, we should be better equipped to understand what does it mean to build an ethical algorithm. We have seen that ethics is about both the ends and the means of technology, so that what needs now to be properly understood is how an algorithm can be good or bad. But first, what is an algorithm anyway?

> At its most fundamental level, an algorithm is nothing more than a very precisely specified series of instructions for performing some concrete task. The simplest algorithms—the ones we teach to our first-year computer science students—do very basic but often important things, such as sorting a list of numbers from smallest to largest (Kearns & Roth, 2020).

This task can be carried out in different ways, so that we can have "bad" algorithms when we have a set of instructions that achieve the goal, but they do it in a longer way than others can do. So, bad in this case would mean not fast enough. This is not the meaning covered by Algorethics. An algorithm is good or bad depending on its design, that is on how much attention has been paid by the developers to make sure that values like privacy, anonymity, accountability, transparency, traceability, responsibility, reliability, trustworthiness, and many more have been addressed and embedded as far as possible in the design. This list, as we said, is far from complete, almost every day we discover ethical areas that are involved in the algorithms reach and that deserve to be adjusted, or more broadly to be taken in consideration, so not to damage anyone or anything.

Broadly speaking, we could say that the task of Algorethics is the effort to keep the man in the loop, to quote a Benanti's book recently released (Benanti, 2022). The main goal of this kind of ethics is to make sure that the decision-making process remains within the human grasp. We can go further than this, adding that algorithms, differently from other tools, take decisions, develop decisional models, for this reason we need to understand how these decisions affect others lives and how to pay attention to the outcomes. Using an algorithm is not like using a hammer, of course we can do bad things using hammers, but this does not depend on the hammer itself, or its design, it is a very specific tool, projected for very specific tasks, even if we can use it not properly. Algorithms have a degree of autonomy, they don't just serve a purpose, they are more flexible and can be adopted for different purposes. This greater flexibility, together with their "agency" (Arvan, 2022) is what makes them more morally delicate to adopt.

Let's make an example of this view. Let's say we would like the algorithm predicting loan repayment not to be racially biased. It seems just fair. But, if the accuracy of the algorithm model depends also on this bias, once we got it cleaned from the bias, the result will necessarily be a less accurate model. It looks like that the right thing to do comes at the cost of a less efficient performance.

What Algorethics puts in evidence is that computation is not a neutral thing. Someone might think that, after all, the job of algorithms is nothing but a procedural thing, detached from any moral connotation, but this evaluation would be quite misleading, since the computational part of the algorithm is a step of broader decision process, whereas taking decisions is by any means a moral task. It is a logical mistake to fail in seeing how the cognitive aspect of a decision process is essential to the deliberation itself, so how we acquire knowledge, matters, how the algorithm provide us certain information, matters, how the data are selected, processed, and eventually deployed matters.

Algorethics in short helps us to recognize the value of the reasoning process we need to go through while deliberating. The fact that today a large part of this knowing process is located in machines is the new landscape we need to address, but it's not a completely new chapter of the human morality. The means according to which we now can carry out this task might be new ones.

The question that now opens to us is: how can human beings flourish or thrive in such new environments? How is it possible to adopt these new technologies without being trapped in their alienating nature? This is what we will try to address in the next two paragraphs.

6 On-Life

Today in 2023, a person who asked us whether we are *online* or *offline* would be prompted to answer that the question is from the 1990s, since precisely the experience of being connected and networked has become so pervasive as to render the question itself meaningless. In fact, the term *on-life* was coined to describe this state of affairs, to hold together both the fact of being systematically *online* and that of continuing to live in the biological environments that have always characterized the human sphere. *Onlife*, again, to say in some way that human nature has always also been artificial nature and that the specifics of this artificiality today are declined in its being in connection with others, with machines or artificial intelligences.

As already mentioned, Luciano Floridi, the Author who coined the term in early 2013, explains the new concept as «the new experience of a hyperconnected reality within which it is no longer sensible to ask whether one may be online or offline» (Floridi, 2015). Reference is thus made to a new experience of hyper-connected reality, that is, of a reality in which the distinction between *online* and *offline* ceases to be meaningful or, to put it better, the *offline* no longer exists while the ubiquity of connections turns out to be an inescapable fact. This new scenario evidently does not leave the surrounding ecology without consequence; not because the world was not connected before, it was, it always has been evidently, to paraphrase a famous author, the flap of a butterfly's wings (Bradbury, 1952) could always cause a Tsunami thousands of miles away, but the kind of connection we are exposed to today is different and opens the world to new scenarios that need to be analyzed.

Floridi, when proceeding to analyze this new reality, articulates his examination in five points:

1. the blurring of the distinction between reality and virtuality;
2. the blurring of the distinction between human, machine and nature;
3. the reversal from information scarcity to information abundance; and
4. the shift from the primacy of stand-alone things, properties, and binary relations,
5. to the primacy of interactions, processes and networks (Floridi, 2015).

The key term to describe this process of homogenization between *online* and *life* is that of *blurring*. *Onlife* is characterized by the blurring of the distinction between reality and virtuality, as well as that between man, machine and nature. There is also the inversion of the model of information scarcity in favor of that of information abundance, if not overabundance, as well as the re-hierarchizing of objects in favor of the primacy of the interaction of processes and networks. In other words, according to Floridi, there is a shift from the primacy of substance to that of relation/connection.[3]

To use an image adopted by the author in several of his speeches, one can think of the plant phenomenon of mangroves,[4] which are special plants that live in brackish waters on the border between the salty waters of the seas and the fresh waters of the rivers, in a mixed ecosystem where precisely the boundary between the salty and the non-salty is blurred. Let us now try to understand these descriptive lines in more depth.

The first: blurring between reality and virtuality. It is generally thought that the virtual is precisely the non-real, so that between the two dimensions there would be an unbridgeable gap of sorts. In truth, and in this Floridi has many reasons, the boundary between the two realms is much more porous and blurred than it appears. Virtual in fact has never designated the not real, but rather the potential, what could be there, not simply what is not there. So that in effect reality always carries a load of possibility, of virtuality if you will, undeniably. Having premised this, however, we must then specify that the sense in which *onlife* is taken would obliterate the difference between real and virtual is that the virtualities offered by computer connection would always be active. That is, in some way we are proposed a world where the virtual is always, constantly, present no longer only as potentiality, but as

[3] A further step forward in this direction is observable in (Ball, 2022).

[4] On the idea of new environments is useful to see (Fabris, 2020).

a concomitant operative reality. In other words, the world of the possible is represented under the aspect of the concrete present, the living. *Onlife*.

This shift from the possible to the present, actual and contemporary is not without consequences. The first among these is precisely the oblivion of the dimension of possibility, crushed on the pole of ubiquitous activity. Everything seems just a click away, space and time somehow compress, giving the impression that, once online, these limitations vanish. This impression somehow of omnipotence, or at any rate of perpetual ubiquity and synchronicity, is precisely apparent. Space and time continue to configure our essential limits and possibilities and thinking of them as unoptimized impediments can risk losing a sense of reality.

The second blurring that *onlife* would produce is that between man, machine, and nature. Again, it should be remembered that man has always been *homo faber*, as well as *sapiens*, and that machines therefore have always configured a natural landscape in its basic ecology. That then today's machine civilization is very different from the pre-modern environment is certainly true, but this truth does not refute the fact that the technical-operational capacity of the human is inherent in us. The difference that *onlife* thus brings is not so much to be seen in the fact that today the machines to which we are connected are somehow necessary and ubiquitous, as already noted, but rather that these machines, far from being thought of as prostheses of human potential, begin to rise to the rank of intrinsic ends of human action. That is, there is in some ways a dangerous inversion of means with ends. It is humans who are prostheses of the machine and work for it: think of the generation of *big data* for example in the field of social media, where the currency of exchange for the use of such platforms is precisely the consent to use the data generated by users, who thus become the real operators of the system, before being users. And all this is perceived as natural, that is, as entirely consistent with the architecture of the digital system in which we are immersed.

Now, while it is natural that machines are in our genetic code, it is less so that they constitute an end in themselves. In this sense, the example of the latest generation of chatbots like ChatGPT is clarifying. AI machines that improve with constant use, so that the end of their use becomes in some way the use itself, configuring that circularity/fluidity half-ending that was noted just above.

Lastly, the shift/inversion from the model of informational scarcity to that of superabundance, which entails among other things the eclipse of objects understood in their own substantiality in favor of a processuality that would precede the objective rigidity of things. According to classical philosophical vocabulary we find ourselves here in the paradigm of becoming at the expense of that of being, Heraclitus in place of Parmenides. Somehow this inversion is the original downstream at the beginning of all those already pointed out in the previous points. Said differently, the world of blurring blurs the contours of its objects in favor of the dynamic that invests them, so that there are no longer stable things, but figures in motion where the boundary between what is moving and the environment in which the displacement takes place, still, is imprecise, necessarily changing. The problem is no longer that of the scarcity or absence of information, but the opposite problem of its incalculable overabundance, so that the researcher tending to identify the missing aspect of phenomena, now becomes a discriminator tending to differentiate background noise from the soundtrack. In environments such as that of the information overabundance, it is necessary to discriminate what is important, relevant, from what is not. Differentiating becomes the most important cognitive skill.

To conclude, it is perhaps useful to note that these new scenarios in truth always reflect the specifics of the digital world, that is, of that environment dominated by the 0–1 logic typical of computer writing. This cognitive model is discrete, 0–1 precisely, true or false, admitting no gradualness, much less blurring. It is therefore curious that in such a model, Cartesian we might add, the new "fuzzy" *onlife* paradigm finds its place. Yet, this is a fruitful paradox, in fact it allows us to see how the digital rigidity of the *aut aut*, either this or that, *tertium non datur*, then fails to remain true to itself. The linearity of the digital, *online*, when it expands to the world of human relationships, pervading them and making itself an instrument of them, betrays itself by blurring its own boundaries. The blurring can be seen as a symptom of the original excessive rigidity that today calls for rethinking in order to integrate into the flexible human sphere by definition (Pessina, 2023).

That is, it is necessary to equip ourselves with the tool of analogy to understand and practice wisely the new digital *on-life* dimensions. Indeed, the analogical device allows the right degree of composition between rigidity and flexibility, to glimpse the structure, rigid, together with its projection, dynamic, while maintaining the distinction between reality and possibility, which configures the human *proprium*.

7 Digital Human Flourishing

The term *human flourishing* refers to the fullest realization of human life in accordance with its nature and potential: «Flourishing itself might be understood as a state in which all aspects of a person's life are good. We might also refer to such a state as complete human well-being, which is again arguably a broader concept than psychological well-being» (VanderWeele, 2017). Interestingly enough, to describe the human drive towards full self-realization, the verb used—*to flourish*—belongs to the field of botany. The flourishing of a plant depends on environmental conditions—conditions that are strictly spatio-temporal. On one hand, humidity, soil composition and quality, exposure to sunlight, weather conditions, and temperature—in short, *space*. On the other hand, the length and succession of day and night, the changing of seasons, and the phases of plant growth: *time*. The plant is not the only one responding to space and time: non-human animals are also rooted—another interesting word derived from botany—in their spatio-temporal environment. The animal is always facing something or someone, it is always rooted in the here and now. According to the philosopher Max Scheler, unlike non-human animals who always say yes to reality, humans are eternal protesters against simple reality, *those who can say no*, ascetics of life (Scheler, 2009). The ability to overcome its biological constraints opens human life to possibilities unattainable by any other animal: humans can surpass the limits imposed by their biology and surrounding environment, achieving a degree of freedom unparalleled in nature. But it is worth asking whether, as the flip side of the coin, freedom from biological and environmental constraints prevents humans from achieving full realization in the digital environment, which we have seen is detached from space and time. *Is it possible to flourish in a digital environment if the first condition of flourishing is to be embedded in time and space, to be present in material reality?* This is the key question. To flourish in a digital environment seems to require a reappropriation of space and time. Is it perhaps too provocative to hypothesize a return to our biological nature, to our kinship with the plant and animal world, and to the spatio-temporal rootedness that characterizes it?

In recent years, various apps have emerged to regulate time spent on digital platforms. An initiative of similar inspiration is the "Day of Unplugging," in which people are invited to disconnect from technological devices for 24 h. The purpose is to encourage a digital break, allowing people to reconnect with themselves, with others, and with the surrounding environment by engaging in offline activities. Both apps and the day of unplugging are attempts to redirect people to real space-time. One of these apps is called *Forest*, and its slogan reads 'stay focused, be present.' The user accesses the application when needing to focus, *to be present*: by keeping the app open—thereby resisting the temptation to access other browsing windows—they will see a tree grow on the screen. If they exit the app, the tree will 'die'. What strikes about the *Forest* app is its naturalistic mold: trees, living beings rooted by constitution, become emblems of presence. The paradox of *Forest* is that it conceives presence as an achievement snatched from an environment that, as we have seen, lacks the necessary characteristics to be truly present. These forms of time and space control assume that the situation of normality—which as such is never questioned—is digital. Disengagement is conceived as a pause, as an extra-ordinary initiative—literally, "out of the ordinary"—and is pursued with an inevitable return in mind:

> The very apps that people use to help them with digital disengagement in turn lead to a process where 'real life' time is found only by adjusting digital time – where the digital is the default and starting point – rather than adjusting digital time based on real lifetime. […] This digi-centric logic means that our perception of time and our temporal organization of everyday life is structured by the very architecture of the platforms used (Kuntsman & Miyake, 2022).

The individual may also seek to reclaim their own time and space, but their operation remains an end in itself because the state of normalcy is digital: the rest of the world waits, or rather demands, their return to this state. In this sense, society is complicit in the digital-centric setting of individuals' experiences: «Digitality is constant. As such, opt-out cannot be constant if we are to be part of a society that collectively conforms, perpetuates, and is complicit in maintaining a 'starting point' that is necessarily digital» (Kuntsman & Miyake, 2022).

The digital is the numerical projection of a real environment imagined differently: it is the numerical formalization of a reality that is deeply embodied and woven with space-time. In this formalization, the material dimension is lost, and it is this dimension that must be reclaimed. If the starting point returns to be reality, it is possible to project its spatio-temporal categories in the digital environments: «All that is solid melts into air and clocks ooze a different, non-millenarian time. This is iTime but it can be clocked, must be clocked, in the modernist way, with downtime as well as uptime, night and morning, boundaries» (Agger, 2011). In iTime *all that is solid melts*: for digital human flourishing, it is necessary to reverse the process of liquefaction and return to the solidity of reality—a reality that is *rigidly spatio-temporal*. For authentic digital human flourishing, one must *root* oneself, put down roots in reality. Let's take a look now to the anthropological side of the question, that we may call the human ecology.

8 Towards a Human Ecology[5]

The Oxford Dictionary defines ecology as «the branch of biology that deals with the relationships between living organisms and their environment». *Digital* ecology, in particular, focuses on the interactions between humans, as well as between humans and digital technologies, in digital environments. If ecology *deals with the relationships between living organisms and their environment*, it may well be dealing with how digital environments interact and influence human behaviours, social relationships, and well-being.

According to Marshall McLuhan, every technology medium *shapes an environment*. With the word 'environment', he refers not only to the physical context in which we live but also the social and cultural context that surrounds us. Each medium doesn't only transmit contents, but also—and mainly—modifies the context in with these contents are received and interpreted. Media are never 'neutral' instruments, but rather environments that actively influence our culture, society, and ways of living. With the saying 'the medium is the message' McLuhan expresses the idea that the fundamental message that a medium conveys is its own existence and its peculiar ways to influence our experience and perception of reality. Given that inherent power of digitality, we must ask ourselves: which kind of environments does it shape? Do these environments *enhance* or *damage* human well-being? More broadly: *are digital environments suitable for human flourishing or not?*

As already pointed out, in digital environments our perception of time and space is altered. One emblematic example is the change that working and leisure time are undergoing in the post-modern era. With the spreading of telework, the definition of working time and spaces has significantly blurred, almost like the distinction between being online or offline that brought to the new onlife era. The 'flexibility' allowed by this new form of work demands job performances outside the working time—a time that loosens its boundaries: if the worker is 'always available', always 'online' in the digital environment, he is potentially able to work anytime,

[5] The topic shows a growing literature Cf. (Geetha Devi, 2019; Park, 1936; Valera, 2023). Useful insights also in (Naess, 2008)

anywhere. «One of the implications of the digital reality is that the once clear and fixed boundaries of time and place have become blurry and amorphous» (Katsabian, 2020): the case of telework clearly demonstrates this point. To tackle this issue, it would be wrong to focus on the specific ICTM that render telework possible: as McLuhan points out, is *the environment* that these technologies create that must be considered, not the technologies *in themselves*.

Some scholars use the concept of 'time porosity' to define the phenomenon of overlapping temporalities: «Time porosity refers to the contemporary forms of interference between working time and personal time, transcending the traditional opposition between work and non-work» (Genin, 2016). But the overlapping of working time and personal time is a peculiar characteristic of an equally peculiar environment, the digital one. The environment generated by digital technologies has given rise, in relation to work, to new production models involving heavy workloads and the extended availability for work. The implications of this new environment have become the subject of interest for the European Union, which has begun to talk about the *right to disconnect*, or «the right of workers not to engage in work-related activities or communications by means of digital tools (including inter alia smartphones, tablets, laptops, and desktop computers) or other communication tools outside working time»[6] (Dagnino, 2021). In a 2021 Resolution, the European Parliament «highlights that constant connectivity combined with high job demands and the rising expectation that workers are reachable at any time can negatively affect workers' fundamental rights, their work-life balance, and their physical and mental health and well-being».[7] Being constantly connected, everywhere and at any time, undermines workers' well-being: if the conditions of full presence through the boundaries of space and time are not guaranteed, the well-being of the individual is threatened.

It is expected that the European Union may further examine the issue of disconnection and may consider the introduction of regulations or guidelines at the community level to ensure workers' right to disconnect throughout the EU. The point is precisely this: that of disconnection *is not an individual but a collective issue*. The individual has little hope of being able to regain full presence if inserted in a digital environment that is constitutively unable to guarantee it. This is the hypocrisy of applications like *Forest*: they encourage the individual to make a titanic effort, that of disconnecting, destined to fail if the environment from which he tries to disconnect *is structured not to make him disconnect*. The point is to imagine and shape a digital environment that protects the right to the full presence of the individual. To this environment must be placed boundaries similar to those of our real experience: in other words, temporal and spatial constraints.

---

[6] Guiding Principles on Implementing Worker's Right to Disconnect. Report of the European Law Institute. Retrieved May 12, 2024, from https://digital-strategy.ec.europa.eu/en/policies/digital-principles

[7] European Parliament resolution of 21 January 2021 with recommendations to the Commission on the right to disconnect (2019/2181(INL)). Retrieved May 12, 2024, from https://www.europarl.europa.eu/doceo/document/TA-9-2021-0021_EN.html

According to the European Digital Rights and Principles, one of the main themes of the digital rights is putting people and their rights at the centre of the digital transformation: if people are put on the center of digital transformation, it's *their* way of experiencing the world that must be considered and protected in the first place. The digital environment must be shaped *by* humans and *around* humans, aligning with the profoundly human constraints of time and space, not *vice versa*. The attempt to conceive the digital as a possibility of human reality and not as a reality *detached* from it should be one of the most important, collective strives of our times.

References

Agger, B. (2011). iTime: Labor and life in a smartphone era. *Time and Society, 20*(1), 119–136.

Anderson, C. (2008). The end of theory. *Wired*, 16, 106–107. Retrieved February 5, 2024, from https://www.wired.com/2008/06/pb-theory/

Arvan, M. (2022). Varieties of artificial moral agency and the new control problem. *Humana. Mente Journal of Philosophical Studies, 15*(42), 225–256.

Ball, M. (2022). *The Metaverse: And how it will revolutionize everything*. Liveright Publishing Corporation.

Benanti, P. (2017). *Il dataismo come nuova religione*. Retrieved February 5, 2024, from https://www.paolobenanti.com/post/2017/03/24/il-dataismo-come-nuova-religione

Benanti, P. (2022). *Human in the loop. Decisioni umane e intelligenze artificiali*. Mondadori.

Bertolaso, M., Capone, L., & Rodríguez-Lluesma, C. (2022). *Digital humanism. A human-centric approach to digital technologies*. Palgrave Macmillan.

Boczkowski, P. J., & Mitchelstein, E. (2021). *The digital environment. Learn, work and play now*. The MIT Press.

Bradbury, R. (1952). *A sound of thunder*. Collier's.

Brooks, D. (2013). The philosophy of data. *The New York Times*. Retrieved from https://www.nytimes.com/2013/02/05/opinion/brooks-the-philosophy-of-data.html?smid=url-share

Crippa, M., & Girgenti, G. (2024). *Umano troppo umano, esercizi spirituali contro l'intelligenza artificiale*. Piemme.

Dagnino, E. (2021). Il diritto alla disconnessione nell'esperienza contrattual-collettiva italiana, LAVORO, DIRITTI, EUROPA, 1–13.

De Rosnay, J. (1979). *The macroscope, towards a global vision* (R. Edwards, Trans.). Harper & Row, Publishers. Introduction available online, Retrieved March 14, 2024, from http://pespmc1.vub.ac.be/macroscope/default.html (Original work published 1975).

Fabris, A. (2020). New environments: Rethinking ethics in an era of technological changes. In L. Valera & J. C. Castilla (Eds.), *Global changes: Ethics, politics and environment in the contemporary technological world*. Springer.

Floridi, L. (Ed.). (2015). *The Onlife manifesto, being human in a Hyperconnected era*. Springer.

Fuchs, C. (2022). *Digital humanism, a philosophy for 21st century digital society*. Emerald Publishing.

Geetha Devi, T. V. (2019). *Understanding human ecology. Knowledge, ethics and politics*. Routledge.

Genin, E. (2016). Proposal for a theoretical framework for the analysis of time porosity. *International Journal of Comparative Labour Law and Industrial Relations, 32*(3), 280–300.

Katsabian, T. (2020). It's the end of working time as we know it: New challenges to the concept of working time in the digital reality. *McGill Law Journal, 65*(3), 379–419.

Kearns, M., & Roth, A. (2020). *The ethical algorithm, the science of socially aware algorithm design*. Oxford University Press.

Kuntsman, A., & Miyake, E. (2022). *Paradoxes of digital disengagement. In search of the opt-out button*. University of Westminster Press.

Lindberg, S., & Roine, H.-R. (2021). *The ethos of digital environments. Technology, literary theory and philosophy*. Routledge.

Mantini, A. (2022). Technological sustainability and artificial intelligence Algor-ethics. *Sustainability, 14*, 3215. https://doi.org/10.3390/su14063215

Naess, A. (2008). In A. Drengson & B. Devall (Eds.), *Ecology of wisdom, writings by Arne Naess*. Counterpoint.

Park, R. E. (1936). Human ecology. *American Journal of Sociology, 42*(1), 1–15.

Pessina, A. (2023). *L'essere altrove, l'esperienza umana nell'epoca dell'intelligenza artificiale*. Mimesis.

Scheler, M. (2009). *The human place in the cosmos*. Northwestern University Press. (Original work published 1929).

Valera, L. (2023). *Pantheism and ecology, cosmological, philosophical, and theological perspectives*. Springer.

VanderWeele, T. J. (2017, August 1). On the promotion of human flourishing. *PNAS, 114*(31), 8148–8156.

Neuroergonomics for Human-Centered Technological Contexts

Giacinto Barresi

Abstract Enabling technologies constitute key milestones for humankind, expanding the capabilities of individuals and societies. Such a disruptive impact can also shape the perspective of any users beyond their own mind, body, and environment. For instance, the pervasive adoption of Artificial Intelligence (AI) solutions is already radically transforming our skills and our settings. Particularly, embodied AIs can help us by means of robots and smart environments in multiple situations through interactions based on user interfaces and feedback powered by a progressively more and more refined computational modelling of our intentions and conditions in a specific context. Indeed, it is crucially important to enable a machine to correctly depict a setting as a system including itself, a space, and people whose skills, limitations, needs, and goals should define the ongoing tasks within the context. Interestingly, ergonomics provides effective interdisciplinary strategies to face such a challenge, investigating how a human being interacts with other elements of a system (a technological context, in the mentioned cases). Simultaneously, this discipline also has the scope of improving the well-being of the person and the performance of the system itself. Each (physical, cognitive, organizational) branch of ergonomics offers valuable methodologies for studying the space where human-technology interactions happen and show their effects. In particular, this manuscript will focus on the contributions of neuroergonomics, the discipline investigating human factors according to neuroscientific methods. Such an approach can exploit rich data collected, for instance, by means of wearable devices in ecologically valid contexts based on physical and digital (and phygital) spaces populated by humans and smart machines, making the design of the latter truly centered on the person.

Keywords Human factors · Artificial intelligence · Robotics · Neuroergonomics · Value-sensitive design · Smart environments

---

G. Barresi (✉)
Bristol Robotics Laboratory, University of the West of England, Bristol, UK
e-mail: giacinto.barresi@uwe.ac.uk

A. Fabris, S. Belardinelli (eds.), *Digital Environments and Human Relations*,
Human Perspectives in Health Sciences and Technology 150,
https://doi.org/10.1007/978-3-031-76961-0_4

1 Introduction

Each enabling technology that has appeared in history disclosed novel opportunities for human understanding and intervention, moving the boundaries of possibility in the minds and hands of the member of our species (Clark, 2001; Dolezal & DeFalco, 2023; Papagiannis, 2017; Raisamo et al., 2019). Indeed, the process of innovation itself has expanded the horizon of human perception, cognition, and action. This is definitely clear in the current transformative era, marked by extraordinary changes based (for instance) on the pervasive adoption of Artificial Intelligence (AI) in multiple contexts (including the ones of daily life). Thus, it becomes urgent to investigate how novel technologies can extend the range of human intent and grasp on reality in each potential scenario. The aforementioned integration of AI into everyday contexts definitely represents an exemplary milestone in the human journey (Dwivedi et al., 2021; Taj & Zaman, 2022; Augusto, 2007; Gruetzemacher & Whittlestone, 2022). Unlike anything human-made, AI (and any device controlled by it) is characterized by the unique capability to adapt through intelligent behaviours like learning, also facilitating or enabling fruitful interactions with humans. However, such interactions can be positive just when they are tailored on the human capabilities, limitations, needs, and goals in a certain context—and we must be sure that our intelligent machines "understand" the circumstances defining the situation and the system (Han et al., 2023; De Lima & Feijó, 2019; Quintas et al., 2018; Bertolaso et al., 2022; Fabris, 2020).

In this domain, Embodied AIs generally demonstrate their especially impactful roles: controlling robots and smart environments, they can actually affect our experience and conditions by changing our physical context. If an AI can anticipate our needs or interpret our intentions, an embodied AI can extend the application of such computational processes to real settings beyond a traditional user interface on a display, providing us with assistance and augmentation through intuitive and reality-based interactions (based on gestures, for instance). However, they can also be intertwined with virtual systems for exploiting the advantages of the latter in a myriad of hybrid (digital and real) contexts. This physical dimension of AI highlights its ubiquitous presence in the world. However, designing and deploying such pervasive systems in real contexts can be a complex challenge because it introduces further needs and risks (obviously at physical level too). This requires a careful rethinking of the relationships between humans and machines, leading to a person-centered design that considers the context of our actions as a system including human beings and their interactions with other elements (more or less intelligent) of the system itself (Broering et al., 2021; Kim et al., 2020; Chrisley, 2003; Pfeifer & Iida, 2004). For instance, we must ponder how AI technologies rely on computational models that can collect, process, and interpret vast amounts of data from the environment—including humans. In the latter case, the models underlying such technologies should be designed to represent the human perspective for providing an appropriate behaviour in a specific context to pursue a certain goal. Reaching such an achievement would make an AI able to transcend its role as a tool and truly become a collaborator, enhancing human capabilities—one of the goals of Human-Centered AI design (Xu et al., 2023; Shneiderman, 2022; Margetis et al., 2021).

Accordingly, this can be seen as the premise to the augmentation of human capabilities through AI, shaping human-AI interactions as ways to go beyond our cognitive limitations (exactly like exoskeletons can make us overcome our physical limitations, for instance). In practical terms, such a perspective can enable people to (for example) overcome mental impairments, bypass an initial lack of expertise during an emergency, or enhance even the finest abilities of a skillful professional. Furthermore, it is important to analyze these aspects because of their ethical implications, especially when AI systems make the boundaries between our thoughts and the machine activities blur: this is the case, for instance, of affective computing technologies and devices based on brain-computer interfaces. Such ethical implications can be further expanded when we consider how AIs are embodied, especially in smart environments: this situation can generate issues related to privacy in data collection, making the machines (designed to help us through monitoring features) potential trespassers of our personal space. Understanding our context for performing useful and acceptable actions requires intensive and extensive data collection, which becomes even more detailed when a machine must physically interact with the world. Thus, the design of any embodied AI like the ones in robotic devices needs a deep investigation of human needs and values in each situation we may imagine, especially when we aim at achieving effective, efficient, satisfying, accessible, and fair interactions oriented to the human well-being (He et al., 2021; Ozmen Garibay et al., 2023; Doncieux et al., 2022; Bingley et al. 2023; Xu, 2019; Capel, 2023; Gao et al., 2021; Hui et al., 2022; Lecomte, 2023; Ashok et al., 2022). These intelligent technologies are just among the currently most groundbreaking examples of pervasive artificial systems becoming integral components of our mental and physical landscapes, reshaping the cognitive representation of ourselves alongside our own context. Through us as components of a wider system, the AI solutions can play a fundamental (hopefully positive) role in the effort of expressing our own potential in diverse activities across physical and computer-generated settings (we must remember the major trend of the metaverse and all physical-digital—phygital—dimensions of interaction and experience). Making such contexts more human (and humane) depends on our attention to the multifaceted implications of technologies in the different domains of our individual and social existence (Gaggioli et al., 2023; Cerasa et al., 2022; Strathearn & Ma, 2020; Bassyouni & Elhajj, 2021).

Interestingly, we are often referring to contexts and environments as systems and to humans as elements of the latter. Such a perspective perfectly matches the one of a discipline investigating how people interact with other elements of a system, specifically for understanding the individual capabilities and needs and for designing tools and procedures accordingly to improve the human well-being and the system performance. This is the definition of the disciplinary domain of human factors and ergonomics (these two terms are often used as synonyms) (Bentley et al., 2021; Salvendy & Karwowski, 2021; Karwowski, 2006; Dul et al., 2012). This manuscript has been devised to describe how a specific area of ergonomics—neuroergonomics—can advantageously contribute to understand and solve the issues emerging in technological contexts according to a human-centered perspective that encompasses our subjective experience and the inner activities of our nervous system in interaction with the environment (Dehais et al., 2020; Ayaz & Dehais, 2021).

Next section will be dedicated to quickly introducing the domain of human factors and ergonomics with its three main branches.

The subsequent section will discuss the area of neuroergonomics with its specific perspective, presenting the advantages of this approach, and examples of its application—e.g., neuroergonomics in digital contexts, neuroadaptive techniques, inclusive and ethical dimensions in neurodesign.

The conclusions of the chapter will point at the opportunity of considering neuroergonomics as a potential way to face the challenges posed by technological contexts as human-machine systems.

2 Human Factors and Ergonomics

Human factors and ergonomics (HFE) constitute a multidisciplinary field (defined by domains like medicine, psychology, engineering, design, sociology, computer science, neuroscience) focused on studying the interactions of human beings with other elements of a system. It is already crucial to highlight how this statement considers the human beings as elements of a system: a company, a family, a school, and many other contextually specific systems. A system can even be just a human-robot dyad as a team. However, the scope of HFE can be found beyond a typical laboratory investigation: indeed, an ergonomist should employ theory, principles, data, and methods from the disciplines encompassed in this field to design, assess, and re-design the system in order to optimize its overall performance alongside the well-being of its human "elements" in an ecologically valid setting (a situation similar to real ones) or an actual context of activity. The search for safety was probably one of the major drives for the birth and the development of HFE, searching for solutions increasing human safety instead of focusing on productivity: the person's conditions must always be the priority of any ergonomist. Overall, the essence of HFE lies in the multidisciplinary study of human skills, limitations, and needs in a certain context for improving the design of systems in terms of processes and assets. Accordingly, we can say that HFE shapes contexts that are both efficient and comfortable, reducing the likelihood of errors, enhancing safety, and improving productivity: human-centered contexts defined according to criteria like sustainability (Lyons et al., 2023; Guastello, 2023; Karwowski, 2001; Hancock, 2023). The complexity of these challenges is the main reason of the multidisciplinary approaches characterizing HFE. Another consequence of such a complexity is the internal presence of sub-domains within HFE. Indeed, according to the International Ergonomics Association (IEA) and its national societies, the domain of HFE can be divided into three primary branches: physical, cognitive, and organizational ergonomics. Considering the perspectives of such three sub-domains is a wise strategy for initially approaching a problem in terms of aspects that involve the individual, physically and mentally, even within a system increasing the resulting complexity of the interactions occurring in a certain context (and pondering the impact of the activities of the ergonomist even beyond individual use cases) (Dovramadjiev et al., 2023; Helander, 1997; Caple, 2010).

2.1 *Physical Ergonomics*

Physical ergonomics focuses on concerns about human anthropometric, anatomical, physiological, and biomechanical issues. Sometimes it overlaps with the naïve general definition of ergonomics, pondering difficulties caused by a design of tools difficult to grasp or of seats generating bad postures and physical strain. Physical ergonomics mainly aims at optimizing the physical fit between people and objects (e.g., tools) and between people and contexts (e.g., workspaces) in which they operate. An exemplary goal of physical ergonomics is the mitigation of risk of musculoskeletal injuries—for instance, by assessing hand tools to minimize the strain aggravating or generating peripheral neuropathies, or by devising comfortable and adjustable seats with lumbar support for the worker's body (and for different sizes and shapes) to avoid back pain. For instance, an ergonomically designed office chair can reduce back pain and increase comfort for users by providing adequate lumbar support and adjustability to accommodate a wide range of body sizes and shapes. Safety and effectiveness are priorities for physical ergonomics in different contexts, from home to industry to rehabilitation center. Physical ergonomics is especially crucial in contexts powered by robots, considering how, for instance, cobotics (collaborative robotics) is expanding its presence in industrial and healthcare settings: cobots are designed for a safe and productive physical interaction with the human operator, supporting procedures like learning from demonstration to physically program the robot to repeat certain movements (from the manipulation of a product in manufacturing environments, to the mobilization of the limbs of bedded patients) (Lind, 2017; Meidert et al., 2016; Pipher, 2021; Chimera et al., 2021; Goonetilleke & Karwowski, 2017; Jones & Kumar, 2001; Nussbaum & van Dieën, 2006; Fathil et al., 2021; Yan & Jia, 2022; Monari et al., 2024; Rapetti, 2022; Abdulazeem & Hu, 2023; Liao et al., 2023).

2.2 *Cognitive Ergonomics*

Introducing cognitive ergonomics is sometime challenging: this discipline deals with mental processes—e.g., perception, attention, learning, memory, reasoning, planning, decision-making—underlying the human actions during the interaction with other elements of a system, especially through a user interface (UI, the "location" where user and machine "meet") in contexts of Human-Computer Interaction (HCI) and Human-Robot Interaction (HRI). Cognitive ergonomics embraces the assessment and the improvement of the design of UIs for enhancing their usability (with its ISO-based dimensions of efficacy, efficiency, and satisfaction) and reducing the mental workload experienced by the user by means of intuitive solutions generating affordances and constraints. Further examples of cognitive ergonomics to improve human-system interactions: powering the situation awareness of teleoperation users by means of digital cues and alarms, improving the technology acceptance of prosthetic users through multisensory feedback and engaging games for establishing a sense of embodiment. Especially interesting for a cognitive ergonomist is the general domain of user experience: the user's perceptions and reactions resulting from the use and/or anticipated use of a product, service, or system as a whole. Particularly challenging is surely the one of accessibility: in this case, products, services, and systems as a whole (and the activities and experiences depending on them) must be usable by people with temporary, situational, and permanent impairments due to disabilities or other sources of difficulty (e.g., linguistic issues of travelers). Solving these problems help the inclusion of people with diverse backgrounds and abilities in contexts ofwork, education, healthcare, leisure, and many others (family time too). Ais and robotic systems are especially challenging in this domain because, for instance, the first can overload a user interface and the second can have difficulty in establishing intuitive interactions with the user (Barresi et al., 2013; Bitkina et al., 2020; Contreras-Somoza et al., 2021; Tosi, 2020; Banuelos-Lozoya et al., 2021; Grundgeiger et al., 2021; Parasuraman et al., 2008; Sauer et al., 2020; Wenk et al., 2023; Zhang & Hansen, 2022; Kim et al., 2020; Barresi et al., 2021; Caserta et al., 2022; Esentaş, 2021; Tavares et al., 2023; Prewett et al., 2010).

2.3 *Organizational Ergonomics*

Organizational ergonomics—also known as macroergonomics because of its wide perspective on systems—is the branch dedicated to the optimization of socio-technical systems through improvements on teams, management structures, policies, scheduling, safety and security, and multiple systemic processes. Organizational ergonomics embraces social and cultural dynamics in the design of any system encompassing multiple entities intertwined to pursue shared and individual goals. We could say that this branch of ergonomics is devised to improve the system outcome holistically, paying attention not only on the well-being of each person in such a system but also teams. Efficiency and sustainability are not considered by this branch just in terms of productivity but also about the wellness of the members of the organization itself. Ethical issues can become crucial for organizational ergonomics in specific cases, and they should always be checked. Principles like a healthy work-like balance (through flexible work hours and tele-work, for instance) are fundamental like safety promotion strategies, in this domain. Even the improvement of information and communication flow within a context like a company are targeted by organizational ergonomics in order to reach efficacy and satisfaction in the individuals and in their teams. In this domain, especially interesting is the contribution to technological design for solving organizational issues that may be caused by smart systems and robotic partners in teams, as in the context of cobotics, mentioned above (Vink, 2020; Lee et al., 2024a; Firescu, 2021; Privitera, 2020; Koirala & Nepal, 2022; Fadhilah et al., 2024; Carayon, 2021; Lee et al., 2024b; Hendrick & Kleiner, 2002; Hendrick, 1991; Kleiner, 2006; Schmidbauer, 2022; Delgoulet & Santos, 2022; Kant, 2018, 2016; Pinheiro et al., 2021).

Obviously, it is quite difficult at this level keeping strict criteria for separating ergonomics in three branches: teamwork is based on physical and cognitive processes that must follow socio-technical policies. Because of this reason, macroergonomics can offer a framework for connecting multiple sub-domains of human factors: even the most specific ones like environmental ergonomics or architectural ergonomics are transversal in terms of physical, cognitive, and organizational factors. This leads to a need of integration of the three main branches with their further extensions like social ergonomics and affective ergonomics (terms that can be used for focusing on social interactions and emotional-motivational processes that need to be analysed even beyond their relationship with cognitive phenomena) (Fritzsche et al., 2014; Johnsen et al., 2017; Walker et al., 2006; Salas et al., 2008; Nagamachi, 2016; Prastawa et al., 2019; Bannon, 1998; Montague & Asan, 2012; Bornewasser, 2012; Holden et al., 2015).

2.4 *Integration of the Three Branches of HFE*

The integration of physical, cognitive, and organizational ergonomics is crucial in HFE for achieving methodologies leading to holistic improvements of human-technology interactions in terms of well-being of the person (and the team) and performance of the system as a whole. Each branch of ergonomics has specific perspectives and techniques for understanding the human conditions and actions in a certain context. However, such perspectives and techniques do not exclusively belong to just one branch—tools like questionnaires, videoanalysis, SWOT analyses, physiological sensors can be used by experts focusing on the physical, cognitive, and organizational aspects. Thus, such aspects should not be forcefully separated: human well-being and system performance are the targets of the ergonomists and improving them must be the priority (possibly involving colleagues from multiple disciplinary fields for depicting each problem in the most complete way). The complexity of the problems we are investigating can certainly require analytical approaches for obtaining specific findings, but also a synthetic effort can lead to groundbreaking results (Lee et al., 2014; Mehta, 2016; Karsh et al., 2014; Panjaitan et al., 2022; Ehrensberger-Dow & O'Brien, 2015; Kolfschoten & Brazier, 2013).

Interestingly, one of the most recent sub-domains of ergonomics seems offering both approaches to improve our understanding of humans in their activities and interactions with systems and improving the latter accordingly. This is the case of neuroergonomics.

3 Neuroergonomics and Technological Contexts

Neuroergonomics can be considered as the investigation of human factors through methods and techniques of neuroscience, bringing them outside typical laboratory settings for studying the nervous system at work, in real or ecologically valid conditions. In such conditions, the complexity of real contexts can be maintained, making people employ their skills through the same patterns of daily life, without oversimplified situations proposed in laboratory (which can risk losing the subject's compliance to the test procedure, or to not involve their expected set of abilities). Obviously, experimental control must be maintained as well but, luckily, the progressive refinement of our methodologies is also sustained by the amount of heterogeneous big data that can now be collected through wearable and distal sensors. Through this approach, it is possible to collect (statistically) significant results and (subjectively) meaningful information by using the framework offered by specialties like cognitive and affective neuroscience and the tools for physiological and motor data collection that can now be based on low-cost ubiquitous devices like smartwatches and environmental thermal cameras. Obviously, such solutions cannot directly provide information on the brain activity (as it happens through caps for electroencephalography, EEG, or functional infrared spectroscopy, fNIRS) but, as in the approach of peripheral neuroergonomics, they can disclose novel perspectives on their correlates in other biosignals (from heart rate variability, HRV, to galvanic skin response, GSR). Bringing these sensors around, possibly adding eye-tracking glasses for knowing what triggered a certain reaction, it is possible to open new views on neuroscientific research, which receives the findings of neuroergonomists for validating their hypotheses beyond the laboratory limitations. However, it is possible to confuse neuroergonomics with disciplines like psychophysiology because of the techniques that are used. On the other hand, anyway, neuroergonomics is inherently connected to human-centered design (designing systems according to human skills, limitations, and needs in a certain context) in ways that are not included in the scope of psychophysiology, more focused on the investigation of the physiological correlates of mental processes and behaviours regardless of the requests of design. This is especially intriguing because neuroergonomists can explore neurocognitive phenomena—even using tools with lower spatial and temporal resolution (like the portable EEG caps that can be used for controlling video games by classifying one or two brainwave patterns related to meditation and relaxation)—in locations that are difficult to replicate in laboratory (in particular when professionals like pilots or surgeons are working). It is also interesting that, even if neuroergonomics is strictly connected to cognitive ergonomics (it can be considered as a sub-discipline emphasizing the adoption of neuroscientific methods), it also offers powerful insights to the other two branches of ergonomics and receives methodological contributions from them. For instance, problems related to nerve conduction can constitute physical neuroergonomic issues, and employing physiological data for evaluating the stress of workers interacting with a cobot can be seen as a strategy of organizational neuroergonomics (Parasuraman, 2003; Fedota & Parasuraman, 2010; Hancock et al., 2006; Ayaz & Dehais, 2018; Dehais & Callan, 2019; Barresi et al., 2022a; Giorgi et al., 2023; Cassioli et al., 2021; Aljuaid, 2019; Wascher et al., 2023; Barros et al., 2016; Peysakhovich et al., 2019; Bröhl et al., 2017; Del Vecchio et al., 2021; Gramann et al., 2021; Savković et al., 2022a; Baker & Castelli, 2020).

This last point is crucially important, as demonstrated by the examples of applications of neuroergonomics to technological contexts presented below: indeed, a system for monitoring the internal state of a worker and informing a cobot, can create the conditions for making the cobot adapt to the conditions of the person to promote their well-being (especially when stress and fatigue can lead to human errors) and the productivity of the human-machine team. This highlights how neuroergonomics can be considered as a key for both scientific progress and technological advances that can face the challenges posed by systems like the AI mentioned in the first section of this chapter. Furthermore, it can exploit such intelligent systems for refining the analysis of data representing the neural activity of an individual and its cognitive-behavioural correlates in contexts beyond the laboratory (even exploiting more and more portable devices for EEG and fNIRS): neuroergonomics can bring ecologically valid data to understand the context of interaction, and AI systems can power such research process while they achieve neuroergonomic models of contextual representation to (for instance) improve their capability to plan the activity of a robot according to the user's expectations (McKendrick et al., 2016; Zakeri et al., 2022, 2023; Savković et al., 2022b; Caiazzo et al., 2023; Burden et al., 2022; Asgher et al., 2023). Nevertheless, reaching these futuristic scenarios needs further efforts to investigate the individual neurocognitive processes, as next subsections will discuss for introducing opportunities offered by neuroergonomics for analyzing and modeling the contexts of human-machine interaction and for making it appropriate according to human perspectives.

3.1 Neuroergonomics in Digital Contexts

The versatility of such approaches is wide, and it embraces contexts that are not necessarily (totally) "real": virtual reality (VR) and any combination of virtuality and reality (e.g., augmented reality, AR; augmented virtuality, AV; mixed reality, MR) falling in the domain of extended reality (XR). Through such systems, neuroergonomists can simulate ecologically valid settings or they can engage subjects (even patients in rehabilitation) in repetitive activities that are necessary for collecting data. Interestingly, trends like the one of the technologies for the metaverse (assuming that we still do not have any persistent shared digital dimension, but we will probably have a "metaverse of services and products" per each company needing this type of medium) and all ways to blend digital and physical contexts (as in phygital settings) are already envisioned in future evolutions intertwined with neurotechnologies, which lead to further neuroergonomic advancements and reasons for analysis. We can consider a neurotechnology as a device that can monitor or influence the nervous system: in the first case, it can just be based on sensors; in the second case we know that it is affecting its target because literature informed us that a certain activity has such a consequence, or because there is an exchange of energy between the neural substrate and the device itself, which is literally inducing changes in activation thresholds or plasticity as in neuromodulation, for instance. Creating a neuroergonomic continuum between human and machine through neurotechnologies like brain-computer interfaces (BCIs) and brain-robot interfaces (BRIs) can shape novel dimensions of intuitive interaction and truly human-sensitive contexts (making them phygitally reacting to our will and reactions). This could also happen through AI-based solutions like digital human models and digital twins (digital replicas of physical systems, devised for analysing present and future states of the latter in order to intervene timely), potential tools for neuroergonomists trying to understand and predict the conditions of the user of a robotic system (even remotely, as in a "meta-laboratory" mimicking the distant human-machine system). These are just examples of the potential of neuroergonomics in digital (also hybrid as phygital) technological contexts that are not necessarily connected to AI systems, even if there are growing spaces for intertwining these technologies (Stapleton & Davies, 2011; Vasarainen et al., 2021; Parsons et al., 2020; Lupetti et al., 2015; Barresi et al., 2022b, c, 2015; Galimberti et al., 2019; Omarali, 2023; Zhou et al., 2023; Wang, 2023; Hieu et al., 2023; Abdelghafar et al., 2023; Ullah et al., 2023; Aloqaily et al., 2022; Khang et al., 2023; Huynh-The et al., 2023; Pouyakian, 2022).

It must also be noticed how most examples provided by this sub-section and others in this chapter confirm the dual relationship between intelligent technologies and neuroergonomics: the first are both targets and tools for the second. The latter case can include the assessment of individual conditions according to physio-motor data in order to re-design the system generating a sub-optimal performance or negatively affecting the well-being of the user. However, neuroergonomics works also directly on enriching a device with an intelligent behaviour according to the live monitoring and interpretation of the individual interacting with it. This class of cases encompasses adaptive technologies, one of the most characteristic features offered by neuroergonomics.

3.2 *Neuroadaptive Features*

Improving the design of an intelligent interactive system in order to make it human-centered typically works in two ways. It is possible to establish that a certain set of basic features in the user interface make it more intuitive and, in general, more usable (we can refer to interaction paradigms or just colors and fonts). On the other hand, AI solutions enable the possibility to make the device itself smart in actively collecting data and modeling the conditions of the user in order to adapt to them and lead such conditions to a more positive state (in terms of individual well-being and system performance). This could happen automatically or even with an active contribution of the user (as in biofeedback and neurofeedback) for multiple goals: from relaxation to training. Overall, anyway, neuroadaptive technologies—adjusting to users' cognitive and affective states—are excellent examples of this approach of making technologies inherently neuroergonomic in their own functions more than just evaluated through neuroergonomic methods. We can consider, for instance, how this can make the machine provide the person with, for instance:

- personalized assistance—e.g., intelligent technologies can tailor specific support for people with mood disorders, adjusting parameters of light in rooms for stimulating a positive change in affective states (monitored through wearables);
- performance enhancement—e.g., a smart workplace can optimize task parameters to maintain high performance and minimize the risk of errors according to the levels of cognitive load in workers (for instance by changing tasks when people show frustration);
- adaptive learning—e.g., educational or training tools can modify content and pacing to suit learners' conditions (especially the levels of attention), improving engagement and, consequently, acquisition and retention of knowledge and know-how;
- co-adaptive calibration—e.g., a neurointerface can exploit the same type of approach in the previous point (for instance through neurofeedback) for facilitating the brain-computer interfacing by engaging the user and their attention in tasks like motor imagery, making the task as an effort of co-adaptation between the machine (which is adjusting the calibration procedure to ease the task for the person) and the human (which is learning by trials and errors how to produce the expected pattern).

Implementing these features requires advanced biosignal processing and machine learning to interpret (central and peripheral) neural data accurately in terms of individual cognitive and affective states alongside the context and for clarifying the human perspective on the context. The focus, specifically, is more on spontaneous reactions and implicit interaction than the typical paradigms based on voluntary commands of the users towards the machine (Fairclough, 2023; Appriou et al., 2018; Grubov et al., 2024; Armani et al., 2023; Krol et al., 2022; Kosunen et al., 2017; Salminen et al., 2019; Demazure et al., 2021; Andreessen et al., 2021; Billard et al., 2016).

By pondering the last point in the list of examples above, we can also highlight the opportunities offered by co-adaptive features in robot learning processes, especially in domains like assistive robotics and prosthetic bionics.

Typically, robot training is one of the main domains that can employ the results of human factors research to make this process intuitive for the user. A groundbreaking achievement in this field certainly is Learning from Demonstration (LfD): e.g., the human trainer demonstrates what action should be performed by the robot by being observed (matching the models of the bodies of human and machine) or by holding the robot itself and repeating the expected trajectory and action (kinesthetic learning, quite intuitive for the human if compared to the traditional programming by coding). However, typical trial-and-error methods in LfD and in most methodologies of robot learning often struggle due to the limitations in informativeness and promptness of human tutor's feedback, conveyed through physical contacts, gestures, and vocal instructions (a huge limitation if the user has motor impairments). Nevertheless, neuroscientific research offers paradigms that can be implemented in robot trainings for making the human mentor tacitly and spontaneously offer more detailed feedback within a fraction of a second, even before the end of the robot action that should be evaluated to check if the training has been performed correctly. This is the case of paradigms based on the detection of error-related potentials (ErrPs), EEG potentials correlated with the spontaneous reaction of the observer to an event that is violating their expectations. Thus, by asking the human mentor to wear an EEG cap, it would be possible to inform the machine about upcoming mistakes if it follows the ongoing motor plan or choice: this way, we could obtain a Learning from (violation of) Expectation—LfE. LfE empowers robots (and any intelligent device) to adjust their actions to align with their human mentor's expectations, offering a significant shift in robot cognition and human-robot interaction: it could mimic the behaviour of a child looking at the parent to understand from their face if the ongoing action will be correct or not, stopping before its completion if necessary. This model provides the robot planning with the human's ability to foresee the outcomes of these actions (obviously with a very limited generalization capability), fostering a human-centered predictive situation awareness that simplifies the personalization of the machine even with contextual features: indeed, by adding information on the location and the dynamics of the interaction with objects and environments (for instance by connecting eye-tracking and EEG for finding more precisely the reason of the ErrP—e.g., holding a certain part of a fragile piece of furniture) the AI in this careful machine would acquire a certain level of "common sense" according to the context. We can consider these aspects as matter of laboratory research (especially because, in a reinforcement learning procedure, LfE would provide only the "punishment") but they already show promising results for a future technology transfer that would ease the introduction of robots in real contexts. Interestingly, these processes can be part of the research route towards optimal human-machine co-adaptation that is already producing results in fields like prosthetics, especially in the case of bionic limbs. Exemplary cases are offered by studies on interactive machine learning: the AI in this case is used for making the co-learning of human and prosthesis based on approaches inspired by the constructivism, bringing the experiential component of into the loop of contextual co-adaptation (Hettinger et al., 2003; Iwane et al., 2023; Schaal, 1996; Tavassoli et al., 2023; Kim et al., 2017, 2023; Vukelić et al., 2023; Tao et al., 2023; Wang et al., 2022a, b; Bettoni & Castellini, 2021; Nowak et al., 2018; Beckerle et al., 2022).

3.3 *Value-Sensitive and Inclusive Neurodesign*

The example of LfE in the previous section makes clear the potential of connecting neuroergonomic monitoring with AI in order to make a machine learn the human perspective on a certain action in a specific situation: the human point of view is part of the context itself. In particular, it brings a certain type of evaluation and appraisal that is strictly connected to what we consider right or wrong. Certainly such labels are used in a clearly specific way, without specific references to moral aspects. However, we can consider a principle of human-centered design the one of making the machine follows human expectations, which can be directly tied to one's values like the personal well-being. Value-sensitive design (VSD) is an excellent framework for introducing such aspects in neuroergonomics, while the latter can also integrate the first. The need for VSD can be found in the ethical issues of all systems generating significant concerns about privacy because of their intensive collection of data on people, as it happens in most smart ecosystems through environmental and wearable sensors monitoring our routine. Privacy-by-design is the typical approach that can be adopted in such cases but VSD can offer a wider framework to multiple ethical concerns. Its versatility is based on a multidisciplinary perspective working by means of a synergy of conceptual investigations (defining ethical and legal frameworks about values and their potential violations), empirical investigations (studying real-world and ecologically valid human-machine interactions, in particular the behaviour of users and stakeholders), and technical investigations (developing human-centered solutions for avoiding unexpected and unacceptable situations due to the behaviours of users exploiting breaches or unexpected ways to employ a technology). Emphasizing such human-centered design principles is definitely essential in AI technology transfer but it should be a priority even during the first phases of technology transfer, far before market access. Neuroergonomics and VSD have a dual relationship: neuroergonomic technologies can be analysed according to VSD for evaluating ethical concerns raised, for instance, by the monitoring of the nervous system and the interpretation of biosignals; however, neuroergonomics can offer methodologies for empirical and technical studies of VSD, possibly generating insights for conceptual studies on the human being in specific technological contexts. About values, a final observation should be made about the strategies for increasing the trust of people in AI technologies, as in the case of explainable AI (XAI): making an intelligent system able to explain its own functioning and goals surely is advantageous for establishing a positive interaction with humans (especially thanks to the recent boost provided by Large Language Models, LLM, used by solutions like ChatGPT) through messages based on language (as paradigms like natural interaction and reality-based interaction propose). Neuroergonomics can (responsibly) integrate these systems by offering solutions to connect the semantic knowledge on activities and contexts in XAI to the characterization of neurocognitive and behavioural processes in datasets for making a machine able to adopt the proper strategy in a specific situation (Friedman et al., 2017; Rajpura et al., 2023; Kim et al., 2023; Kiani et al., 2022; Petrushin et al., 2022; Friedman, 1996; Winkler & Spiekermann, 2021; Schaar, 2010).

Especially interesting can be the adoption of both neuroergonomics and VSD in contexts requiring the promotion of inclusion, considering the needs of people with disabilities and other sources of diversity. Certainly, neuroergonomics can improve the accessibility of interactive and intelligent technologies by understanding the capabilities a person can employ, despite any impairment, for generating contexts where everyone can contribute on the same level (from universal design of games for people with and without disabilities, to metaverse-oriented solutions to enable people's attendance in online meetings while their differences are hidden). It is possible to imagine a future of neuroinclusive technologies, for instance, by making them monitor the stress of fragile people like elderly to change the levels of sensory or information load on a user interface, or enabling schools and workplaces with devices that, according to neuroscientific research, can be easier to be handled by people with cognitive disabilities (e.g., visual maps can make organizing and presenting ideas easier) during a class or a meeting. Such a distinction of neuroergonomic technologies intentionally mirrors the one described above on neurotechnologies. Caring for the needs of everyone is a way to shape sustainable contexts, making intelligent technologies a source of equity instead of divide. These are directions that the neurodesign (a design explicitly based on neuroscientific discoveries) should follow through neuroergonomics and according to VSD (Larradet et al., 2017; Lyons et al., 2023; Hutson & Hutson, 2024; Ok & Rao, 2019; Zallio, 2023; Tracada & Sorensen Allacci, 2023; Molina et al., 2024; Zignego et al., 2023; von Thienen et al., 2023).

4 Conclusion

Facing the challenges generated by the disruptive impact of enabling technologies like the ones of AI on our lives, this chapter offered an introduction to a comprehensive and versatile way to approach the profound transformations brought by these advances in our individual and social journeys.

The integration of AI and other advanced technologies into daily life and activities is the premise of a dramatic change in our physical and mental capabilities and potential for action: this happened after the introduction of any enabling solution in the history of humankind, but we are witnessing the emergence of novelties due to the capability of partially or fully autonomous artificial systems to anticipate our actions and reactions, interpret our intentions and needs, and intervene (hopefully) to help us. The opportunities and the risks due to the artificial decision-making processes underlying such tasks are just examples of the reasons for understanding how machines can behave and for finding ways to make them "understand" the contexts of their behaviours in order to perform actions appropriate from the point of view of users and stakeholders. Such problems are inherently part of the discussion on the relationships between humans and machines, on their interactions and the systems encompassing them. Ergonomics and human factors, as highlighted in the chapter, constitute an elective multidisciplinary domain for facing the mentioned challenges because of their attention for systems from a human-centered perspective in research and design.

The role of all branches of ergonomics becomes clear if we consider, for instance, the physical, cognitive, and organizational advantages and issues generated by the pervasive adoption of embodied AI in robots and smart environments. In this scenario, ergonomics must provide us with the means for a leap towards realizing artificial agents and settings able to catch the implications of the contexts alongside the human conditions and goals before enacting any intervention. However, this makes even greater the challenge constituted by the development of complex models that mirror our intentions and conditions according to contextual information for making intuitive and productive human-technology synergies. The computational effort of interpreting our subjective and contextual dynamics can be successfully sustained by data on our own physiological and behavioural processes: as proposed in this chapter, neuroergonomics can offer methodologies and approaches for modeling the human-centered technological contexts and, accordingly, enable a refined tuning of the AI functions based on intentions and spontaneous reactions of people.

Specifically, neuroergonomics emerges as a pivotal discipline through its capability to employ physio-motor data collected by means of pervasive (environmental and wearable) sensors, even low-cost ones. This may build and offer an eclectic framework for analysing the multifaceted interactions between humans and technologies within shared contexts. By employing neuroscientific methods and ubiquitous devices outside the laboratory (in ecologically valid or even real settings), neuroergonomics provides insights into the physical, cognitive, and organisational implications of the interactions occurring in human-technology systems for redesigning the latter. This approach paves the way for the development of AI solutions that are inherently human-centered—pondering our perspectives and our roles as humans, and also as users or stakeholders involved in co-design processes.

The chapter also introduced this topic through the description of three sample domains: neuroergonomics in digital contexts (considering the innovations brought by virtual and augmented settings, and perspectives on metaverse, digital twins, phygital interactions with potential synergies with neurotechnologies), neuroadaptive processes (technologies can interpret and anticipate our cognitive and affective changes for providing an appropriate way to adjust the functions of a machine according, for instance, to the violation of our expectations), value-sensitive and inclusive neurodesign (extending the conceptual, empirical, the technical studies in ethically characterised design through the use and for the development of neuroergonomic devices). The systemic approach of ergonomics is certainly maintained in neuroergonomics, which offers both research and design methods for making technologies contextually responsive, enhancing individuals abilities and quality of life while they holistically power the system performance too. We could say that neuroergonomic technologies can complement and integrate our skills and well-being, augmenting our experience without replacing us: they need humans as references (e.g., as sources of "common sense") for working, even just catching and processing signals we produce implicitly.

However, we must remember how these functions do not "spontaneously" emerge (yet?) from the AI itself: we must always look for the accountability of the individuals and organisations shaping such systems and placing them on the market with specific goals. The integration of neuroergonomic principles in AI development obviously underscores precise choices with consequences that are not always unpredictable (thus, someone can be accountable). Even if certain enabling technologies seem risky for the human dignity (because of human naivety or malice?), we need anyway to explore their potential for helping us according to our values, making the latter guide the human-centered design processes using or shaping novel AI-powered neuroergonomic solutions.

This approach not only increases the potential benefits of AI but also mitigates the risks associated with its pervasive adoption in technology-powered physical, digital, and phygital spaces (and times, considering the twinning strategies to explore future states of a system) for making them truly person-sensitive and, through this, context-aware.

References

Abdelghafar, S., Ezzat, D., Darwish, A., & Hassanien, A. E. (2023). Metaverse for brain computer Interface: Towards new and improved applications. In A. E. Hassanien, A. Darwish, & M. Torky (Eds.), *The future of Metaverse in the virtual era and physical world* (pp. 43–58). Springer.

Abdulazeem, N., & Hu, Y. (2023). Human factors considerations for quantifiable human states in physical human–Robot interaction: A literature review. *Sensors, 23*(17), 7381.

Aljuaid, A. M. (Ed.). (2019). Theoretical design of eeg-based neuroergonomics integrated portable system, applying direct psychophysiological indicators. In *2019 Industrial & Systems Engineering Conference (ISEC)*. IEEE.

Aloqaily, M., Bouachir, O., Karray, F., Al Ridhawi, I., & El Saddik, A. (2022). Integrating digital twin and advanced intelligent technologies to realize the metaverse. *IEEE Consumer Electronics Magazine*.

Andreessen, L. M., Gerjets, P., Meurers, D., & Zander, T. O. (2021). Toward neuroadaptive support technologies for improving digital reading: A passive BCI-based assessment of mental workload imposed by text difficulty and presentation speed during reading. *User Modeling and User-Adapted Interaction, 31*, 75–104.

Appriou, A., Cichocki, A., & Lotte, F. (2018). Towards robust neuroadaptive HCI: exploring modern machine learning methods to estimate mental workload from EEG signals. In *Extended Abstracts of the 2018* CHI conference on human factors in computing systems. ACM.

Armani, F., Daly, I., Vernitski, A., Gillmeister, H., & Scherer, R. (2023). Maths anxiety and cognitive state monitoring for neuroadaptive learning systems using electroencephalography. In *2023 IEEE International conference on metrology for eXtended reality, artificial intelligence and neural engineering (MetroXRAINE)*. IEEE.

Asgher, U., Ayaz, Y., & Taiar, R. (2023). Advances in artificial intelligence (AI) in brain computer interface (BCI) and industry 4.0 for human machine interaction (HMI). *Frontiers in Human Neuroscience, 17*.

Ashok, M., Madan, R., Joha, A., & Sivarajah, U. (2022). Ethical framework for artificial intelligence and digital technologies. *International Journal of Information Management, 62*, 102433.

Augusto, J. C. (2007). Ambient intelligence: The confluence of ubiquitous/pervasive computing and artificial intelligence. In A. J. Schuster (Ed.), *Intelligent computing everywhere* (pp. 213–234). Springer.

Ayaz, H., & Dehais, F. (2018). *Neuroergonomics: The brain at work and in everyday life*. Academic.

Ayaz, H., & Dehais, F. (2021). Neuroergonomics. In G. Salvendy & W. Karwowski (Eds.), *Handbook of human factors and ergonomics* (pp. 816–841). Wiley.

Baker, B., & Castelli, D. (2020). Physical activity and sedentary behavior influences on executive function in daily living. *Neuroergonomics: Principles and Practice*, 161–181.

Bannon, L. J. (1998). CSCW: Towards a social ergonomics? *The RTO human factors and medicine panel symposium held in Edinburgh, United Kingdom, 20-22 April 1998*, k2.1-k2.10.

Banuelos-Lozoya, E., Gonzalez-Serna, G., Gonzalez-Franco, N., Fragoso-Diaz, O., & Castro-Sanchez, N. (2021). A systematic review for cognitive state-based QoE/UX evaluation. *Sensors, 21*(10), 3439.

Barresi, G., Deshpande, N., Mattos, L.S., Brogni, A., Guastini, L., Peretti, G., et al. (Eds.). (2013). Comparative usability and performance evaluation of surgeon interfaces in laser phonomicrosurgery. In *2013 IEEE/RSJ international conference on intelligent robots and systems*. IEEE.

Barresi, G., Olivieri, E., Caldwell, D. G., & Mattos, L. S. (2015). Brain-controlled AR feedback design for user's training in surgical HRI. In *2015 IEEE International conference on systems, man, and cybernetics*. IEEE.

Barresi, G., Marinelli, A., Caserta, G., De Zambotti, M., Tessadori, J., Angioletti, L., et al. (2021). Exploring the embodiment of a virtual hand in a spatially augmented respiratory biofeedback setting. *Frontiers in Neurorobotics, 15*, 683653.

Barresi, G., Caldwell, D.G., & Mattos, L.S. (Eds.). (2022a). Natural surgeon interfaces: Perspectives and examples of intuitive laser control systems in the μRALP project. *Congress of the Italian Society of ergonomics and human factors*. Springer.

Barresi, G., Pacchierotti, C., Laffranchi, M., & De Michieli, L. (2022b). Beyond digital twins: Phygital twins for neuroergonomics in human-robot interaction. *Frontiers in Neurorobotics, 16*, 913605.

Barresi, G., Zenzeri, J., Tessadori, J., Laffranchi, M., Semprini, M., & De Michieli, L. (2022c). Neuro-gerontechnologies: Applications and opportunities. In S. Scataglini, S. Imbesi, & G. Marques (Eds.), *Internet of things for human-centered design: Application to elderly health-care* (pp. 123–153). Springer Nature Singapore Pte Ltd.

Barros, R. Q., Tavares, A. S., Albuquerque, W., da Silva, J. C., de Lemos, I. A., de Albuquerque Cardoso, R. L. S. et al. (2016 July 17–22). Analysis of product use by means of eye tracking and EEG: A study of neuroergonomics. In A. Marcus (Ed.), Design, user experience, and usability: Novel user experiences: 5th international conference, DUXU 2016, held as part of HCI international 2016. Proceedings, part II (5). Springer.

Bassyouni, Z., & Elhajj, I. H. (2021). Augmented reality meets artificial intelligence in robotics: A systematic review. *Frontiers in Robotics and AI, 8*, 724798.

Beckerle, P., Castellini, C., Lenggenhager, B., & Dosen, S. (2022). Embodiment and co-adaptation through human-machine interfaces: At the border of robotics, Neuroscience and Psychology. *Frontiers in Neurorobotics, 16*, 871785.

Bentley, T., Green, N., Tappin, D., & Haslam, R. (2021). State of science: The future of work–ergonomics and human factors contributions to the field. *Ergonomics, 64*(4), 427–439.

Bertolaso, M., Capone, L., & Rodríguez-Lluesma, C. (2022). *Digital humanism: A human-centric approach to digital technologies*. Palgrave Macmillan.

Bettoni, M. C., & Castellini, C. (2021). Interaction in assistive robotics: A radical constructivist design framework. *Frontiers in Neurorobotics, 15*, 675657.

Billard, A. G., Calinon, S., & Dillmann, R. (2016). Learning from humans. In B. Siciliano & O. Khatib (Eds.), *Springer handbook of robotics* (pp. 1995–2014). Springer.

Bitkina, O. V., Kim, H. K., & Park, J. (2020). Usability and user experience of medical devices: An overview of the current state, analysis methodologies, and future challenges. *International Journal of Industrial Ergonomics, 76*, 102932.

Bornewasser, M. (2012). Social and organizational ergonomics and temporary work in industrial firms. In P. Vink (Ed.), *Advances in social and organizational factors* (pp. 379–389). CRC Press.

Broering, A., Niedermeier, C., Olaru, I., Schopp, U., Telschig, K., & Villnow, M. (2021). Toward embodied intelligence: Smart things on the rise. *Computer, 54*(7), 57–68.

Bröhl, C., Theis, S., Rasche, P., Wille, M., Mertens, A., & Schlick, C. M. (2017). Neuroergonomic analysis of perihand space: Effects of hand proximity on eye-tracking measures and performance in a visual search task. *Behaviour & Information Technology, 36*(7), 737–744.

Burden, A. G., Caldwell, G. A., & Guertler, M. R. (2022). Towards human–robot collaboration in construction: Current cobot trends and forecasts. *Construction Robotics, 6*(3), 209–220.

Caiazzo, C., Savkovic, M., Pusica, M., Milojevic, D., Leva, M. C., & Djapan, M. (2023). Development of a neuroergonomic assessment for the evaluation of mental workload in an industrial human–robot interaction assembly task: A comparative case study. *Machines, 11*(11), 995.

Capel, T., & Brereton, M. (2023). What is human-centered about human-centered AI? A map of the research landscape. In A. Schmidt & K. Väänänen (Eds.), *Proceedings of the 2023 CHI conference on human factors in computing systems*. ACM.

Caple, D. C. (2010). The IEA contribution to the transition of ergonomics from research to practice. *Applied Ergonomics, 41*(6), 731–737.

Carayon, P. (2021). Social and organizational foundation of ergonomics: Multi-level systems approaches. In G. Salvendy & W. Karwowski (Eds.), *Handbook of human factors and ergonomics* (pp. 227–235). Wiley.

Caserta, G., Boccardo, N., Freddolini, M., Barresi, G., Marinelli, A., Canepa, M., et al. (2022). Benefits of the Cybathlon 2020 experience for a prosthetic hand user: A case study on the Hannes system. *Journal of Neuroengineering and Rehabilitation, 19*(1), 68.

Cassioli, F., Fronda, G., & Balconi, M. (2021). Human–co-bot interaction and neuroergonomics: Co-botic vs. robotic systems. *Frontiers in Robotics and AI, 8*, 659319.

Cerasa, A., Gaggioli, A., Marino, F., Riva, G., & Pioggia, G. (2022). The promise of the metaverse in mental health: The new era of MEDverse. *Heliyon, 8*(11), e11762.

Chimera, N. J., Holmes, M. W., & Gabriel, D. A. (2021). Anthropometrics and electromyography as predictors for maximal voluntary isometric wrist torque: Considerations for ergonomists. *Applied Ergonomics, 97*, 103496.

Chrisley, R. (2003). Embodied artificial intelligence. *Artificial intelligence, 149*(1), 131–150.

Clark, A. (Ed.) (2001). Natural-born cyborgs? In *International conference on cognitive technology*. Springer.

Contreras-Somoza, L. M., Irazoki, E., Toribio-Guzmán, J. M., de la Torre-Díez, I., Diaz-Baquero, A. A., Parra-Vidales, E., et al. (2021). Usability and user experience of cognitive intervention technologies for elderly people with MCI or dementia: A systematic review. *Frontiers in Psychology, 12*, 636116.

De Lima, E. S., & Feijó, B. (2019). Artificial intelligence in human-robot interaction. In H. Ayanoglu & E. Duarte (Eds.), *Emotional design in human-robot interaction: Theory, methods and applications* (pp. 187–199). Springer.

Dehais, F., & Callan, D. (2019). A neuroergonomics approach to human performance in aviation. In M. A. Vidulich & P. S. Tsang (Eds.), *Improving aviation performance through applying engineering psychology: Advances in aviation psychology, Volume 3* (pp. 123–141). CRC Press.

Dehais, F., Karwowski, W., & Ayaz, H. (2020). Brain at work and in everyday life as the next frontier: Grand field challenges for neuroergonomics. *Frontiers Media SA, 583733*.

Del Vecchio, A., Castellini, C., & Beckerle, P. (2021). Peripheral neuroergonomics–an elegant way to improve human-robot interaction? *Frontiers in Neurorobotics, 15*, 691508.

Delgoulet, C., & Santos, M. (2022). Ergonomic work analysis and training: Past, present and future. *Work, 73*(s1), S141–SS52.

Demazure, T., Karran, A., Léger, P.-M., Labonte-LeMoyne, E., Sénécal, S., Fredette, M., et al. (2021). Enhancing sustained attention: A pilot study on the integration of a brain-computer interface with an enterprise information system. *Business & Information Systems Engineering, 63*(6), 653–668.

Dolezal, L., & DeFalco, A. (2023). The Posthuman. In A. Hall (Ed.), *Contemporary literature and the body: A critical introduction* (pp. 215–237). Bloomsbury Publishing.

Doncieux, S., Chatila, R., Straube, S., & Kirchner, F. (2022). Human-centered AI and robotics. *AI Perspectives, 4*(1), 1.

Dovramadjiev, T., Dobreva, D., & Zlateva, R. (2023). Advanced application of ergonomics and human factors in academic and scientific activities. *Ergonomics, 7*(1), 49–56.

Dul, J., Bruder, R., Buckle, P., Carayon, P., Falzon, P., Marras, W. S., et al. (2012). A strategy for human factors/ergonomics: Developing the discipline and profession. *Ergonomics, 55*(4), 377–395.

Ehrensberger-Dow, M., & O'Brien, S. (2015). Ergonomics of the translation workplace: Potential for cognitive friction. *Translation Spaces, 4*(1), 98–118.

Esentaş, M. (2021). A leisure time educational tool: Mind and intelligence games: A leisure time educational tool. *International Journal of Curriculum and Instruction, 13*(2), 1355–1373.

Fabris, A. (2020). New environments: Rethinking ethics in an era of technological changes. In L. Valera & J. C. Castilla (Eds.), *Global changes: Ethics, politics and environment in the contemporary technological world* (pp. 9–16). Springer.

Fadhilah, N., Matondang, A. R., & Anizar, A. (2024). Assessing work system processes using macroergonomic analysis and design approach: A literature review. *Jurnal Sistem Teknik Industri, 26*(1), 47–56.

Fairclough, S. (2023). Neuroadaptive technology and the self: A postphenomenological perspective. *Philosophy & Technology, 36*(2), 30.

Fathil, S. M., Ghani, M. R. A., Chen, K.-Y., Lee, P., Hou, J.-D., & Lin, J.-A. (2021). Physical ergonomics in peripheral nerve block. *Applied Sciences, 11*(20), 9736.

Fedota, J. R., & Parasuraman, R. (2010). Neuroergonomics and human error. *Theoretical Issues in Ergonomics Science, 11*(50), 402–421.

Firescu, V. (2021). Ergonomics, sustainability and performance. *Proceedings of the 15th International management conference*, Faculty of Management, Academy of Economic Studies, *15*(1), November, 813–821.

Friedman, B. (1996). Value-sensitive design. *Interactions, 3*(6), 16–23.

Friedman, B., Hendry, D. G., & Borning, A. (2017). A survey of value sensitive design methods. *Foundations and trends® in Human–Computer Interaction, 11*(2), 63–125.

Fritzsche, L., Wegge, J., Schmauder, M., Kliegel, M., & Schmidt, K.-H. (2014). Good ergonomics and team diversity reduce absenteeism and errors in car manufacturing. *Ergonomics, 57*(2), 148–161.

Gaggioli, A., Cerasa, A., & Barresi, G. (2023). Phygital mental health: Opportunities and challenges. In S. Scataglini, S. Imbesi, & G. Marques (Eds.), *mHealth and human-centered design towards enhanced health, care, and Well-being* (pp. 21–35). Springer Singapore.

Galimberti, C., Gaggioli, A., Fabbri, T., Brivio, E., & Vergine, I. (2019). Introducing and implementing phygital and augmented reality at work. *Studi organizzativi, XXI*(2), 137–163.

Gao, X., Wang, Y., Chen, X., & Gao, S. (2021). Interface, interaction, and intelligence in generalized brain–computer interfaces. *Trends in Cognitive Sciences, 25*(8), 671–684.

Giorgi, A., Ronca, V., Vozzi, A., Aricò, P., Borghini, G., Tamborra, L., et al. (2023). Multimodal characterization of mental fatigue on professional drivers. *AHFE International, 102*, 93–102.

Goonetilleke, R.S., & Karwowski, W. (Eds.). (2017). Advances in physical ergonomics and human factors. In *Proceedings of the AHFE 2017 conference on physical ergonomics and human factors*. Springer.

Gramann, K., McKendrick, R., Baldwin, C., Roy, R. N., Jeunet, C., Mehta, R. K., et al. (2021). Grand field challenges for cognitive neuroergonomics in the coming decade. *Frontiers Media SA, 643969*.

Grubov, V. V., Khramova, M. V., Goman, S., Badarin, A. A., Kurkin, S. A., Andrikov, D. A., et al. (2024). *Open-loop neuroadaptive system for enhancing student's cognitive abilities in learning* (Vol. 12, pp. 49034–49049). IEEE Access.

Gruetzemacher, R., & Whittlestone, J. (2022). The transformative potential of artificial intelligence. *Futures, 135*, 102884.

Grundgeiger, T., Hurtienne, J., & Happel, O. (2021). Why and how to approach user experience in safety-critical domains: The example of health care. *Human Factors, 63*(5), 821–832.

Guastello, S. J. (2023). *Human factors engineering and ergonomics: A systems approach*. CRC Press.

Han, X., Huang, D., Eun-Lee, S., & Hoon-Yang, J. (2023). Artificial intelligence-oriented user interface design and human behavior recognition based on human–computer nature interaction. *International Journal of Humanoid Robotics, 20*(06), 2250020.

Hancock, P. (2023). Quintessential solutions to existential problems: How human factors and ergonomics can and should address the imminent challenges of our times. *Human Factors., 66*, 1657–1668. https://doi.org/10.1177/00187208231162448

Hancock, P. A., Szalma, J. L., Parasuraman, R., & Rizzo, M. (2006). Stress and neuroergonomics. In R. Parasuraman & M. Rizzo (Eds.), *Neuroergonomics: The brain at work* (pp. 195–206). Oxford University Press.

He, H., Gray, J., Cangelosi, A., Meng, Q., McGinnity, T. M., & Mehnen, J. (2021). The challenges and opportunities of human-centered AI for trustworthy robots and autonomous systems. *IEEE Transactions on Cognitive and Developmental Systems, 14*(4), 1398–1412.

Helander, M. (1997). Forty years of IEA: Some reflections on the evolution of ergonomics. *Ergonomics, 40*(10), 952–961.

Hendrick, H. W. (1991). Ergonomics in organizational design and management. *Ergonomics, 34*(6), 743–756.

Hendrick, H. W., & Kleiner, B. M. (Eds.). (2002). *Macroergonomics: Theory, methods, and applications*. Lawrence Erlbaum Associates Publishers. https://doi.org/10.1201/b12477

Hettinger, L. J., Branco, P., Encarnacao, L. M., & Bonato, P. (2003). Neuroadaptive technologies: Applying neuroergonomics to the design of advanced interfaces. *Theoretical Issues in Ergonomics Science, 4*(1–2), 220–237.

Hieu, N. Q., Hoang, D. T., Nguyen, D. N., & Dutkiewicz, E. (2023). Toward BCI-enabled Metaverse: A joint learning and resource allocation approach. In *GLOBECOM 2023–2023 IEEE global communications conference*. IEEE.

Holden, R. J., Rivera, A. J., & Carayon, P. (2015). Occupational macroergonomics: Principles, scope, value, and methods. *IIE Trans Occup, 3*(1), 1–8. https://doi.org/10.1080/21577323.2015.1027638

Hui, H., Lin, F., Yang, L., Gong, C., Xu, H., Han, Z., et al. (2022). Affective computing model with impulse control in internet of things based on affective robotics. *IEEE Internet of Things Journal, 9*(21), 20815–20832.

Hutson, J., & Hutson, P. (2024). *Inclusive smart museums: Engaging neurodiverse audiences and enhancing cultural heritage*. Springer.

Huynh-The, T., Pham, Q.-V., Pham, X.-Q., Nguyen, T. T., Han, Z., & Kim, D.-S. (2023). Artificial intelligence for the metaverse: A survey. *Engineering Applications of Artificial Intelligence, 117*, 105581.

Iwane, F., Sobolewski, A., Chavarriaga, R., & Millán, J. R. (2023). EEG error-related potentials encode magnitude of errors and individual perceptual thresholds. *Iscience, 26*(9), 107524.

Johnsen, S. O., Kilskar, S. S., & Fossum, K. R. (2017). Missing focus on human factors–organizational and cognitive ergonomics–in the safety management for the petroleum industry. *Proceedings of the institution of mechanical engineers, part O: Journal of risk and reliability*, 231(4), 400–410. Sage.

Jones, T., & Kumar, S. (2001). Physical ergonomics in low-back pain prevention. *Journal of Occupational Rehabilitation, 11*, 309–319.

Kant, V. (2016). Cyber-physical systems as sociotechnical systems: A view towards human–technology interaction. *Cyber-Physical Systems, 2*(1–4), 75–109.

Kant, V. (2018). The sociotechnical constitution of cognitive work analysis: Roles, affordances and malfunctions. *Theoretical Issues in Ergonomics Science, 19*(2), 195–212.

Karsh, B.-T., Waterson, P., & Holden, R. J. (2014). Crossing levels in systems ergonomics: A framework to support 'mesoergonomic' inquiry. *Applied Ergonomics, 45*(1), 45–54.

Karwowski, W. (2001). *International encyclopedia of ergonomics and human factors*. CRC Press.

Karwowski, W. (2006). The discipline of ergonomics and human factors. In G. Salvendy (Ed.), *Handbook of human factors and ergonomics* (pp. 1–31). Wiley.

Khang, A., Shah, V., & Rani, S. (2023). *Handbook of research on AI-based technologies and applications in the era of the Metaverse*. IGI Global.

Kiani, M., Andreu-Perez, J., Hagras, H., Rigato, S., & Filippetti, M. L. (2022). Towards understanding human functional brain development with explainable artificial intelligence: Challenges and perspectives. *IEEE Computational Intelligence Magazine, 17*(1), 16–33.

Kim, S. K., Kirchner, E. A., Stefes, A., & Kirchner, F. (2017). Intrinsic interactive reinforcement learning–using error-related potentials for real world human-robot interaction. *Scientific Reports, 7*(1), 1–16.

Kim, K., de Melo, C. M., Norouzi, N., Bruder, G., & Welch, G. F. (Eds.). (2020). Reducing task load with an embodied intelligent virtual assistant for improved performance in collaborative decision making. In *2020 IEEE conference on virtual reality and 3D user interfaces (VR)*. IEEE.

Kim, S., Choo, S., Park, D., Park, H., Nam, C. S., Jung, J.-Y., et al. (2023). Designing an XAI interface for BCI experts: A contextual design for pragmatic explanation interface based on domain knowledge in a specific context. *International Journal of Human-Computer Studies, 174*, 103009.

Kleiner, B. M. (2006). Macroergonomics: Analysis and design of work systems. *Applied Ergonomics, 37*(1), 81–89.

Koirala, R., & Nepal, A. (2022). A literature review on ergonomics, ergonomics practices, and employee performance. *Management, 4*(2), 273–288.

Kolfschoten, G. L., & Brazier, F. M. (2013). Cognitive load in collaboration: Convergence. *Group Decision and Negotiation, 22*, 975–996.

Kosunen, I., Ruonala, A., Salminen, M., Järvelä, S., Ravaja, N., & Jacucci, G. (2017). Neuroadaptive meditation in the real world. In *Proceedings of the 2017 ACM workshop on an application-oriented approach to BCI out of the laboratory*.

Krol, L. R., Klaproth, O. W., Vernaleken, C., Russwinkel, N., & Zander, T. O. (2022). Towards neuroadaptive modeling: Assessing the cognitive states of pilots through passive brain-computer interfacing. In S. H. Fairclough & T. O. Zander (Eds.), *Current research in Neuroadaptive technology* (pp. 59–73). Elsevier.

Larradet, F., Barresi, G., & Mattos, L. S. (2017). Effects of galvanic skin response feedback on user experience in gaze-controlled gaming: A pilot study. In *2017 39th annual International conference of the IEEE engineering in medicine and biology society (EMBC)*. IEEE.

Lecomte, P. (2023). Umwelt as the foundation of an ethics of smart environments. *Humanities and Social Sciences Communications, 10*(1), 1–12.

Lee, G. I., Lee, M. R., Clanton, T., Sutton, E., Park, A. E., & Marohn, M. R. (2014). Comparative assessment of physical and cognitive ergonomics associated with robotic and traditional laparoscopic surgeries. *Surgical Endoscopy, 28*, 456–465.

Lee, J., Lin, J.-H., Bao, S. S., & Robertson, M. M. (2024a). Work from home: Facilitators for an effective ergonomic work system. *Applied Ergonomics, 118*, 104276.

Lee, K.-S., Lee, J., & Hwang, J. (2024b). Research trends in ergonomics, industrial safety and health: Semantic network and topic analyses. *International Journal of Occupational Safety and Ergonomics, 30*(1), 20–32.

Liao, Z., Lorenzini, M., Leonori, M., Zhao, F., Jiang, G., & Ajoudani, A. (2023). *An ergo-interactive framework for human-robot collaboration via learning from demonstration* (pp. 1–8). IEEE Robotics and Automation Letters.

Lind, C. (2017). *Assessment and design of industrial manual handling to reduce physical ergonomics hazards—Use and development of assessment tools*. KTH Royal Institute of Technology.

Lupetti, M. L., Piumatti, G., & Rossetto, F. (2015). Phygital play HRI in a new gaming scenario. In *2015 7th International conference on intelligent Technologies for Interactive Entertainment (INTETAIN)*. IEEE.

Lyons, J. B., Hobbs, K., Rogers, S., & Clouse, S. H. (2023). Responsible (use of) AI. *Frontiers in Neuroergonomics, 4*.

Margetis, G., Ntoa, S., Antona, M., & Stephanidis, C. (2021). Human-centered design of artificial intelligence. In G. Salvendy & W. Karwowski (Eds.), *Handbook of human factors and ergonomics* (pp. 1085–1106). Wiley.

McKendrick, R., Parasuraman, R., Murtza, R., Formwalt, A., Baccus, W., Paczynski, M., et al. (2016). Into the wild: Neuroergonomic differentiation of hand-held and augmented reality wearable displays during outdoor navigation with functional near infrared spectroscopy. *Frontiers in Human Neuroscience, 10*, 171788.

Mehta, R. K. (2016). Integrating physical and cognitive ergonomics. *IIE Transactions on Occupational Ergonomics and Human Factors, 4*(2–3), 83–87. https://doi.org/10.1080/21577323.2016.1207475

Meidert, U., Neumann, S., Ehrensberger-Dow, M., & Becker, H. (2016). Physical ergonomics at translators' workplaces: Findings from ergonomic workplace assessments and interviews. ILCEA Revue de l'Institut des langues et cultures d'Europe, Amérique, Afrique, Asie et Australie, 27.

Molina, A. I., Arroyo, Y., Lacave, C., Redondo, M. A., Bravo, C., & Ortega, M. (2024). Eye tracking-based evaluation of accessible and usable interactive systems: Tool set of guidelines and methodological issues. *Universal Access in the Information Society*, 1–24.

Monari, E., Avallone, G., Valori, M., Agostini, L., Chen, Y., Palazzi, E., et al. (2024). Physical ergonomics monitoring in human–robot collaboration: A standard-based approach for hand-guiding applications. *Machines, 12*(4), 231.

Montague, E., & Asan, O. (2012). Considering social ergonomics: The effects of HIT on interpersonal relationships between patients and clinicians. *Work, 41*(Supplement 1), 4479–4483.

Nagamachi, M. (2016). *Kansei/affective engineering*. CRC Press.

Nowak, M., Castellini, C., & Massironi, C. (2018). Applying radical constructivism to machine learning: A pilot study in assistive robotics. *Constructivist Foundations, 13*(2), 250–262.

Nussbaum, M. A., & van Dieën, J. H. (2006). Physical ergonomics. In M. Kutz (Ed.), *Materials and mechanical design* (pp. 762–781). Wiley.

Ok, M. W., & Rao, K. (2019). Digital tools for the inclusive classroom: Google chrome as assistive and instructional technology. *Journal of Special Education Technology, 34*(3), 204–211.

Omarali, P. S. P. (2023). Transcending the metaverse and the metaphygital: An early review and trial of metaverse technologies. In AIP conference proceedings, 2968(1). AIP Publishing.

Ozmen Garibay, O., Winslow, B., Andolina, S., Antona, M., Bodenschatz, A., Coursaris, C., et al. (2023). Six human-centered artificial intelligence grand challenges. *International Journal of Human–Computer Interaction, 39*(3), 391–437.

Panjaitan, N., Hasnida, A.-S., & Ali, A. Y. (2022). Mesoergonomics, the missing part of the macroergonomics and microergonomics approach. *Journal of Engineering, Design and Technology, 20*(3), 649–670.

Papagiannis, H. (2017). *Augmented human: How technology is shaping the new reality*. O'Reilly Media, Inc.

Parasuraman, R. (2003). Neuroergonomics: Research and practice. *Theoretical Issues in Ergonomics Science, 4*(1–2), 5–20.

Parasuraman, R., Sheridan, T. B., & Wickens, C. D. (2008). Situation awareness, mental workload, and trust in automation: Viable, empirically supported cognitive engineering constructs. *Journal of Cognitive Engineering and Decision Making, 2*(2), 140–160.

Parsons, T. D., Gaggioli, A., & Riva, G. (2020). Extended reality for the clinical, affective, and social neurosciences. *Brain Sciences, 10*(12), 922.

Petrushin, A., Freddolini, M., Barresi, G., Bustreo, M., Laffranchi, M., Del Bue, A., et al. (2022). IoT-powered monitoring systems for geriatric healthcare: Overview. In S. Scataglini, S. Imbesi, & G. Marques (Eds.), *Internet of things for human-centered design: Application to elderly healthcare* (pp. 99–122). Springer Nature Singapore Pte Ltd.

Peysakhovich, V., Dehais, F., & Duchowski, A. T. (2019). Why is eye tracking an essential part of neuroergonomics? In H. Ayaz & F. Dehais (Eds.), *Neuroergonomics: The brain at work and in everyday life* (pp. 27–30). Elsevier.

Pfeifer, R., & Iida, F. (2004). Embodied artificial intelligence: Trends and challenges. In F. Iida, R. Pfeifer, L. Steels, & Y. Kuniyoshi (Eds.), *Embodied artificial intelligence* (*Lecture notes in computer science*) (Vol. 3139, pp. 1–26). Springer. https://doi.org/10.1007/978-3-540-27833-7_1

Pinheiro, S., Correia Simões, A., Pinto, A., Van Acker, B. B., Bombeke, K., Romero, D., et al. (2021). Ergonomics and safety in the design of industrial collaborative robotics: A systematic literature review. *Occupational and Environmental Safety and Health, III*, 465–478.

Pipher, Z. (2021). Kinematics and muscle activity of the upper extremity while performing cleaning tasks.

Pouyakian, M. (2022). Cybergonomics: Proposing and justification of a new name for the ergonomics of industry 4.0 technologies. *Frontiers in Public Health, 10*, 1012985.

Prastawa, H., Ciptomulyono, U., Laksono-Singgih, M., & Hartono, M. (2019). The effect of cognitive and affective aspects on usability. *Theoretical Issues in Ergonomics Science, 20*(4), 507–531.

Prewett, M. S., Johnson, R. C., Saboe, K. N., Elliott, L. R., & Coovert, M. D. (2010). Managing workload in human–robot interaction: A review of empirical studies. *Computers in Human Behavior, 26*(5), 840–856.

Privitera, M. R. (2020). Human factors/ergonomics (HFE) in leadership and management: Organizational interventions to reduce stress in healthcare delivery. *Health, 12*(9), 1262–1278.

Quintas, J., Martins, G. S., Santos, L., Menezes, P., & Dias, J. (2018). Toward a context-aware human–robot interaction framework based on cognitive development. *IEEE Transactions on Systems, Man, and Cybernetics, Systems, 49*(1), 227–237.

Raisamo, R., Rakkolainen, I., Majaranta, P., Salminen, K., Rantala, J., & Farooq, A. (2019). Human augmentation: Past, present and future. *International Journal of Human-Computer Studies, 131*, 131–143.

Rajpura, P., Cecotti, H., & Meena, Y. K. (2023). Explainable artificial intelligence approaches for brain-computer interfaces: A review and design space. Preprint retrieved from *arXiv:231213033.*

Rapetti, L. (2022). *Ergonomics for physical human-robot collaboration.* The University of Manchester.

Salas, E., Cooke, N. J., & Rosen, M. A. (2008). On teams, teamwork, and team performance: Discoveries and developments. *Human Factors, 50*(3), 540–547.

Salminen, M., Järvelä, S., Ruonala, A., Harjunen, V. J., Hamari, J., Jacucci, G., et al. (2019). Evoking physiological synchrony and empathy using social VR with biofeedback. *IEEE Transactions on Affective Computing, 13*(2), 746–755.

Salvendy, G., & Karwowski, W. (Eds.). (2021). *Handbook of human factors and ergonomics*. Wiley.

Sauer, J., Sonderegger, A., & Schmutz, S. (2020). Usability, user experience and accessibility: Towards an integrative model. *Ergonomics, 63*(10), 1207–1220.

Savković, M., Caiazzo, C., Djapan, M., Vukićević, A. M., Pušica, M., & Mačužić, I. (2022a). Development of modular and adaptive laboratory set-up for neuroergonomic and human-robot interaction research. *Frontiers in Neurorobotics, 16*, 863637.

Savković, M., Mijailovic, N., Caiazzo, C., Djapan, M., & Vukicevic, A. (2022b). Advanced physical ergonomics and Neuroergonomics research on an assembly workstation. In *OESEM. 19th International conference. "man and working environment". Occupational and environmental safety engineering & management* (pp. 351–357).

Schaal, S. (1996). Learning from demonstration. *Advances in Neural Information Processing Systems, 9*.

Schaar, P. (2010). Privacy by design. *Identity in the Information Society, 3*(2), 267–274.

Schmidbauer, C. (2022). *Adaptive task sharing between humans and cobots in assembly processes.* TU Wien.

Shneiderman, B. (2022). *Human-centered AI.* Oxford University Press.

Stapleton, C. B., & Davies, J. R. (2011). Imagination: The third reality to the virtuality continuum. *2011 IEEE International symposium on mixed and augmented reality—Arts, media, and humanities*, 53–60.

Strathearn, C., & Ma, M. (2020). Modelling user preference for embodied artificial intelligence and appearance in realistic humanoid robots. *Informatics, 7*(28), 28. MDPI.

Taj, I., & Zaman, N. (2022). Towards industrial revolution 5.0 and explainable artificial intelligence: Challenges and opportunities. *International Journal of Computing and Digital Systems, 12*(1), 295–320.

Tao, T., Jia, Y., Xu, G., Liang, R., Zhang, Q., Chen, L., et al. (2023). Enhancement of motor imagery training efficiency by an online adaptive training paradigm integrated with error related potential. *Journal of Neural Engineering, 20*(1), 016029.

Tavares, A., Silva, J.L., & Ventura, R. (Eds.). (2023). Physiologically attentive user interface for improved robot teleoperation. In *Proceedings of the 28th international conference on intelligent user interfaces (IUI'23)*, 776–789. Association for Computing Machinery. https://doi.org/10.1145/3581641.3584084.

Tavassoli, M., Katyara, S., Pozzi, M., Deshpande, N., Caldwell, D. G., & Prattichizzo, D. (2023). Learning skills from demonstrations: A trend from motion primitives to experience abstraction. *IEEE Transactions on Cognitive and Developmental Systems.*

Tosi, F. (2020). Cognitive aspects in user experience design: From perception to emotions. In F. Tosi (Ed.), *Design for Ergonomics* (pp. 281–294). Springer.

Tracada, E., & Sorensen Allacci, M. (2023). Evaluating design features to support inclusive, self-directed, and active healthy living behaviours. *Frontiers in Built Environment, 9*, 1232248.

Ullah, H., Manickam, S., Obaidat, M., Laghari, S. U. A., & Uddin, M. (2023). *Exploring the potential of metaverse technology in healthcare: Applications, challenges, and future directions* (Vol. 11, pp. 69686–69707). IEEE Access.

Vasarainen, M., Paavola, S., & Vetoshkina, L. (2021). A systematic literature review on extended reality: Virtual, augmented and mixed reality in working life. *International Journal of Virtual Reality, 21*(2), 1–28.

Vink, P. (Ed.). (2020). Advances in social and organizational factors. *AHFE International.* https://doi.org/10.54941/ahfe1001594

von Thienen, J., Kolodny, O., & Meinel, C. (2023). Neurodesign: The biology, psychology, and engineering of creative thinking and innovation. In N. Rezaei (Ed.), *Brain, decision making and mental health* (pp. 617–659). Springer.

Vukelić, M., Bui, M., Vorreuther, A., & Lingelbach, K. (2023). Combining brain-computer interfaces with deep reinforcement learning for robot training: A feasibility study in a simulation environment. *Frontiers in Neuroergonomics, 4*, 1274730.

Walker, G. H., Gibson, H., Stanton, N. A., Baber, C., Salmon, P., & Green, D. (2006). Event analysis of systemic teamwork (EAST): A novel integration of ergonomics methods to analyse C4i activity. *Ergonomics, 49*(12–13), 1345–1369.

Wang, Y. (2023). The effectiveness of brain-computer Interface Technology in the Metaverse. *Journal of Artificial Intelligence Practice, 6*(7), 13–22.

Wang, X., Chen, H.-T., & Lin, C.-T. (2022a). Error-related potential-based shared autonomy via deep recurrent reinforcement learning. *Journal of Neural Engineering, 19*(6), 066023.

Wang, X., Chen, H.-T., Wang, Y.-K., & Lin, C.-T. (2022b). Implicit robot control using error-related potential-based brain–computer interface. *IEEE Transactions on Cognitive and Developmental Systems, 15*(1), 198–209.

Wascher, E., Reiser, J., Rinkenauer, G., Larrá, M., Dreger, F. A., Schneider, D., et al. (2023). Neuroergonomics on the go: An evaluation of the potential of mobile EEG for workplace assessment and design. *Human Factors, 65*(1), 86–106.

Wenk, N., Penalver-Andres, J., Buetler, K. A., Nef, T., Müri, R. M., & Marchal-Crespo, L. (2023). Effect of immersive visualization technologies on cognitive load, motivation, usability, and embodiment. *Virtual Reality, 27*(1), 307–331.

Winkler, T., & Spiekermann, S. (2021). Twenty years of value sensitive design: A review of methodological practices in VSD projects. *Ethics and Information Technology, 23*, 17–21.

Xu, W. (2019). Toward human-centered AI: A perspective from human-computer interaction. *Interactions, 26*(4), 42–46.

Xu, W., Dainoff, M. J., Ge, L., & Gao, Z. (2023). Transitioning to human interaction with AI systems: New challenges and opportunities for HCI professionals to enable human-centered AI. *International Journal of Human–Computer Interaction, 39*(3), 494–518.

Yan, Y., & Jia, Y. (2022). A review on human comfort factors, measurements, and improvements in human–robot collaboration. *Sensors, 22*(19), 7431.

Zakeri, Z., Khalid, A., Omurtag, A., Breedon, P., Hilliard, G. (2022). Building trust and safety correlates for autonomous systems using physiological, behavioral, and subjective measures. In Asgher, U. (Ed.), *Industrial cognitive ergonomics and engineering psychology: Proceedings of the 13th AHFE International conference on industrial cognitive ergonomics and engineering psychology.* AHFE International. Retrieved September 5, 2024, from https://doi.org/10.54941/ahfe1001595

Zakeri, Z., Arif, A., Omurtag, A., Breedon, P., & Khalid, A. (2023). Multimodal assessment of cognitive workload using neural, subjective and Behavioural measures in smart factory settings. *Sensors, 23*(21), 8926.

Zallio, M. (2023). *Human-centered Metaverse and digital environments*. AHFE International.

Zhang, G., & Hansen, J. P. (2022). Telepresence robots for people with special needs: A systematic review. *International Journal of Human–Computer Interaction, 38*(17), 1651–1667.

Zhou, H., Lv, H., Wang, R., Wu, H., & Yang, G. (2023). Revitalizing Human-Robot Interaction: Phygital Twin Driven Robot Avatar for China–Sweden Teleoperation. *Chinese Journal of Mechanical Engineering, 36*(1), 124.

Zignego, M. I., Bertirotti, A., Gemelli, P. A., & Pagani, L. (2023). Neurodesign: A game-changer in educational contexts. In Jones, D., Borekci, N., Clemente, V., Corazzo, J., Lotz, N., Nielsen, L.M., & Noel L.A. (Eds.). The 7th International conference for design education researchers, 29 November – 1 December 2023. Retrieved September 5, 2024, from https://doi.org/10.21606/drslxd.2023.014

AI Regulation in Europe: Exploring the Artificial Intelligence Act

Carlo Casonato and Giulia Olivato

Abstract Artificial Intelligence is currently governed by a complex framework of ethical guidelines and scattered sector-specific legislation. Against the backdrop of rapid technological development, the legal landscape is also facing a period of fast paced innovation, and the European AI Act is at the forefront of the regulatory race. However, while there seems to be a general consensus on the opportunity to provide safeguards and legal certainty through comprehensive legal frameworks, regulating such a dynamic and pervasive technology is certainly a highly challenging task for any legislator. The first part of the chapter examines the characteristics (the three Ts) that are necessary for any AI law to be sustainable, future-proof and tailored to the specific technology. Therefore, a law should be Tuned (developed in accordance with scientific expertise and societal pluralism), Timely (able to provide certainty, yet flexible to technological change) and Tailored (focused on the particular technology at hand). Thus, a preliminary analysis of the AI Act is carried out by comparing it with the above features. The second part of the chapter examines analytically the content of the proposed regulation, focusing first on the risk-based approach of the AI Act, its scope and the definition of AI system. It then focuses on an overview of the different levels of risk (including the ad hoc legal framework for general-purpose models) and, in particular, the regulation of high-risk systems. Finally, it

---

The authors made equal contributions to the Introduction and Concluding remarks sections of the article. Additionally, Carlo Casonato authored Part I, and Giulia Olivato authored Part II.

C. Casonato (✉)
Comparative Constitutional Law, Jean Monnet Chair on EU Law and AI, Law School, University of Trento, Trento, Italy
e-mail: carlo.casonato@unitn.it

G. Olivato
Law School, University of Trento, Trento, Italy
e-mail: giulia.olivato@unitn.it

A. Fabris, S. Belardinelli (eds.), *Digital Environments and Human Relations*, Human Perspectives in Health Sciences and Technology 150,
https://doi.org/10.1007/978-3-031-76961-0_5

presents a general framework for regulatory sandboxes, governance and enforcement measures. Considerations on the regulation's ability to remain future-proof and preserve innovation while safeguarding fundamental rights are presented in the Concluding Remarks.

Keywords EU law · Artificial intelligence act · Innovation governance and regulation · Artificial intelligence · Digital transition

1 Introduction

Until recently, the field of artificial intelligence (hereinafter AI) was regulated by only a few norms. If there were many documents indicating ethical principles that could guide the use of AI systems (Fjeld et al., 2020; Floridi et al., 2018),[1] not many binding and enforceable regulatory acts were in force. And even when there were properly binding legal acts, they lacked comprehensiveness, mainly focusing on privacy issues.

This situation of limited legal regulation has over time attracted the attention of many experts, not only jurists but also computer scientists for example, and has been criticized even by many representatives of big tech companies who have called for AI regulation.[2] However, while it seems that the desire for regulation in the sector is strongly shared, opinions regarding the characteristics and contents of such regulatory instruments are highly divided.

In the European Union, this division was overcome through a lengthy process of consensus-building, in which the three main bodies involved in the legislative process (the Commission, the Council, and the Parliament, collectively known as the Trilogue) have reached an agreement on a specific and wide-ranging text.

This chapter is divided into two parts. The first part will synthesize the characteristics that AI law should have to regulate a sensitive, pervasive, and transformative area such as AI in an effective and fair way; and will consider whether the text of the AI Act meets these requirements. The second part will examine the main contents of the AI Act itself.

2 What Law for AI?

When delving into AI law, we emphasize the necessity for it to be grounded in scientific sustainability, future-proof, and tailored to the unique characteristics of the regulated techniques. Beyond the generality and vagueness of these formulations, it

[1] Before the publication of the first version of the AI Act proposal by the European Commission (April 2021), there were already many ethical documents on the subject.

[2] See, for example, the hearings involving heads and CEOs of tech companies (including Meta, Google, OpenAI, Nvidia, and IBM) before the US Senate in September 2023.

is possible to identify certain features that can help make this aspiration concrete and verifiable. We will discuss these features in terms of the three Ts that AI law should possess—the three characteristics for a Tuned, Timely, and Tailored law.

2.1 A Tuned Law (Science and Society)

In the first place, it is important to remember that AI is a highly complex subject, hardly understandable without specific training, and even less governable without adequate and careful scientific preparation and expertise. In the impossibility of turning politicians who are called to regulate it (in particular, members of legislative bodies) into computer science experts, it seems instead necessary that they be assisted in making political decisions by bodies that can provide them with impartial and accurate information on the technical characteristics of the object they intend to regulate. In other words, it seems essential to establish and activate channels of a virtuous political-science contamination that allow legislators to be provided with the appropriate data to create a law that correspond to the specific technical characteristics of AI: a sustainable law, or an evidence-based law.

Comparative law provides some practical examples of how this can be achieved. Generally speaking, for example, the British Parliamentary Office for Science and Technology (POST), since 1989, publishes and makes available to Westminster and its committees scientifically reliable, impartial, and up-to-date studies on topics ranging from biology to health, from energy to the environment, from physical sciences to computer science.[3] In Spain, since 2021, the Oficina de Ciencia y Tecnología is responsible for providing the Congreso de los Diputados with reliable reports on scientific matters.[4] And in Europe, the Scientific Advice Mechanism is tasked with providing independent scientific evidence and policy recommendations to the European institutions by request of the College of Commissioners.[5]

In addition to these permanent bodies, many states usually establish temporary ad hoc committees tasked with providing scientific consultations when the legislature decides to enact laws on particularly complex subjects. Among many examples, one well-known case involved the United States when modifying its legal definition of death. This adjustment stemmed from a proposal put forward at the scientific level by a Harvard Medical School committee (Casonato, 2023). Similarly, regulations concerning Assisted Reproductive Technology (ART) reflect in many countries a decision-making process typically involving scientific committees. For instance, the 1990 British Human Fertilisation and Embryology Act was formulated based on the findings outlined in the Warnock Report in 1984, and the Spanish

[3] See https://post.parliament.uk (accessed May 12, 2024).

[4] See https://www.congreso.es/es/cem/oficinacienciatecnologia (accessed May 12, 2024).

[5] See https://scientificadvice.eu (accessed May 12, 2024).

legislation extensively incorporated recommendations from the Comisión Palacios and the Comisión Nacional de Reproducción Humana Asistida (Penasa, 2015).

Besides being complex from a scientific point of view, subjects like AI also are very sensitive and controverse and exhibit a high degree of social divisiveness. According to some, for example, AI is a kind of panacea, capable of solving all our major problems; according to others, on the contrary, AI risks being our last invention, spelling the end of humanity itself (Bostrom, 2014; Barrat, 2013; Harari, 2016; O'Neil, 2016; Russell, 2019; Zuboff, 2019; Crawford, 2021).

Typically, parliamentary and ordinary political party frameworks fail to intercept and represent the diverse and articulated positions present in society. Therefore, there is a general need for an additional element of participatory democracy.[6]

Comparative law, once again, offers effective examples in this direction. When the French Parliament addresses highly controversial and divisive issues such as those related to bioethics, it convenes the États Généraux de la Bioéthique, so that the Parliament can legislate not only based on numerous studies, and reports produced by various involved bodies, but also on a necessary preliminary phase of collective reflection, fostering the emergence of an agreement based on the recognition of shared values.[7] Another form of extensive societal consultation was established, also in France, in 2022, with the aim of reflecting on the opportunity to modify end-of-life regulations. It was the Convention Citoyenne sur la fin de vie held by the Conseil économique, social et environnemental (CESE): comprised of one hundred and fifty citizens, randomly selected based on a series of indicators, it aimed at faithfully representing social pluralism and experimenting a particular form of deliberative democracy.[8]

Considering these perspectives, there arises a question as to whether the AI Act has relied upon scientifically accurate information provided by technical-scientific bodies, and whether it has been developed through a decision-making process that genuinely encompasses the extensive societal diversity.

In broad terms, these inquiries tend to yield affirmative responses. The final text approved by the European Parliament in mid-March 2024 and officially published in July 2024 (European Union, 2024) represents the culmination of an extensive and inclusive decision-making process that engaged a diverse array of stakeholders and bodies. The first version of it was published by the Commission (the body representing the EU's overall interests) in April 2021, and already emerged from a

---

[6] Because of the pervasive nature of AI, it is important to guarantee diversity of opinion and experiences and ensure the participation and active contribution of different stakeholders (and most of all civil society) in the policy making process. With regard to broader AI governance, participation of civil society should also be strongly encouraged within standard setting organisations, because of the relevance that standards will have in the AI Act (cfr. Part II of the article).

[7] See https://www.vie-publique.fr/rapport/37417-etats-generaux-de-la-bioethique-rapport-synthese (accessed May 12, 2024).

[8] See https://www.lecese.fr/convention-citoyenne-sur-la-fin-de-vie (accessed May 12, 2024).

preparatory phase involving the consideration of numerous documents and soft law instruments. Notably, this included European Parliament resolutions on the ethical principles of AI, robotics, and related technology, as well as on the liability regime for AI (both dated October 20, 2020), and on the use of AI (January 20, 2021). The Commission's White Paper on Artificial Intelligence (February 19, 2020) was also taken into account, with its overarching principles subject to vigorous discussion during an intensive consultation phase that concluded in May 2020.[9] Furthermore, the April 2021 regulatory proposal was informed by the efforts of the High-Level Expert Group on AI and the guidelines it formulated regarding Trustworthy AI (April 8, 2019).[10] And before its approval, it underwent an impact assessment procedure conducted by the European Commission's Regulatory Scrutiny Board. Subsequently, the text underwent thorough scrutiny and elicited a series of proposed amendments from other EU bodies participating in the regulatory process. Notably, the Council (representing the Member States) reached a consensus on the general position[11] of the EU Member States in December 2022 (Council, 2022), while the Parliament (representing EU citizens) voted on its stance in June 2023 after proposing numerous amendments (European Parliament, 2023).[12]

In line with the ordinary legislative procedure (the 'co-decision' procedure), an agreement was reached in the Trilogue: a series of equally composite tripartite meetings between the European Commission, the Council of the EU, and representatives of the European Parliament. Through this process, the final version, endorsed by the EU Parliament in mid-March (pending final review by lawyer-linguists), emerged as the outcome of a procedure finely tuned to both the technical and scientific dimensions, as well as the diverse societal positions.

In the long run, regulatory sandbox frameworks (instruments used to facilitate the development and testing of innovative AI systems under strict regulatory oversight before market placement: rec. 138 of the AI Act), can also serve to evaluate the effectiveness of AI legislation in regulating the sector. This evaluation ensures alignment with European values without impeding the sector's development.

---

[9] See https://commission.europa.eu/publications/white-paper-artificial-intelligence-european-approach-excellence-and-trust_en (accessed May 12, 2024).

[10] See https://digital-strategy.ec.europa.eu/en/library/ethics-guidelines-trustworthy-ai (accessed May 12, 2024).

[11] A general position is a political agreement reached by the Council before the Parliament votes on the text for the first time.

[12] The Parliament proposed the extension of prohibited practices, the strengthening of the role of supervisory bodies and the provision of a fundamental rights impact assessment. The Parliament's proposal also set some «General principles applicable to all AI systems», which were not included in the final version of the text.

2.2 A Timely Law

A second characteristic that defines AI as a subject of law is the speed of its evolution. Since its origin in the 1950s, AI has undergone rapid development, with continuous «summers and winters», with new programs and systems entering the market at a steady pace (Russell & Norvig, 2021). With recent strong computational power, vast amount of available data (Big Data), and substantial investments, particularly from private entities, AI is currently experiencing a phase akin to a "tropical summer," epitomized by the launch (November 2022) and widespread adoption (100 million users in a few months) of ChatGPT-3. The continuous release of new models and the rapid pace of change in AI system features are additional aspects that the law must carefully consider when seeking to regulate AI. Otherwise, there is a risk that regulatory efforts may be immediately thwarted by the creation of new and different systems.

The need for the law to be «stable yet not still» is not a new concept—Roscoe Pound already addressed it a century ago (Pound, 1923). Accordingly, comparative law offers a range of tools in this regard, effectively balancing the necessary certainty and predictability of the law with its equally important flexibility and adaptability. The use of sunset clauses or resorting to experimental legislation, for instance, can be cited as approaches unified by the need to tie the effectiveness of law to its review and potential adaptation over time (Casonato, 2023).

Within the AI Act, we find some tools designed to meet this important requirement.

The text of the AI Act is supplemented, in the first place, by annexes that are crucial. These annexes serve various purposes, including the identification of the category of high-risk AI systems. For these high-risk systems, specific conformity assessment procedures and particular requirements are envisaged. And many of the provisions within the AI Act are dedicated to addressing these high-risk systems, as we will see in the second part of this Chapter. For the amendment of these annexes, the Commission is empowered under Article 290 TFEU (delegated acts), and assisted by the Comitology Committees (Art. 98). Thus, as the application of the Regulation highlights any applicative problems, changes can be made to the Act Through the modification of the annexes, and outside of the ordinary, time-consuming legislative procedure otherwise required for the formal revision of the Regulation.

Secondly, the AI Act includes a general obligation to review the Regulation 5 years after its entry into force and, subsequently, every 4 years thereafter. In addition, the Commission should carry out further periodic assessment, such as that of the need to amend the list of high-risk AI systems and the list of prohibited practices once a year (recital 174).

This choice is commendable in view of the aforementioned mobility and dynamic nature of the regulated subject matter, allowing for the pursuit of AI law that is 'stable yet not still'. Indeed, it is no coincidence that a similar periodic verification tool is found in Canada, for example, where the Directive on Automated Decision-Making, entered into force in April 2019, undergoes review every 2 years.[13]

A third mechanism ensuring the adaptability of the AI Act over time involves the guidelines that the Commission may issue to specify practical implementation and enhance the clarity of regulatory articles. These guidelines will, for instance, include a comprehensive range of practical examples covering systems classified as non-high-risk due to exceptions outlined in the AI Act.

The staggered implementation of the Regulation (discussed below) is another instance of the Act's timely approach, as the applicability of provisions pertaining to practices posing unacceptable risks and to the regulation of general purpose AI models, will be enforced ahead of the broader Act.

2.3 *A Tailored Law*

AI is also characterized by not being a singular technology; rather, it encompasses a multitude of systems that exhibit significant variations in technological characteristics, potentials, and risks. For example, when considering the distinctions between model-based systems, machine learning, and neural networks, and the varying levels of power, autonomy, and risks associated with phenomena like the Black Box, the broad spectrum of different approaches within the field of AI becomes evident. From this perspective, it is imperative that the law be finely tailored to address different systems individually, taking into account their respective strengths and weaknesses. An overly general and abstract law would lose its effectiveness, as it would overlook these differences. Therefore, legal frameworks must be crafted with precision to ensure they adequately address the complexities inherent in various AI systems (a Tailored law).

From this perspective, the AI Act emerges as a regulatory model that prioritizes specificity among different systems rather than being a generic law. However, instead of merely distinguishing rules based on different AI systems or areas of application, the EU regulation adopts a risk-based approach. This approach precisely addresses the need for nuanced consideration, as will be discussed in the second part of this chapter. Rather, one may wonder if the AI Act, with one hundred eighty recitals, one hundred thirteen articles, and thirteen annexes, does not constitute a web so dense and intricate as to be sometimes unmanageable.

Considering the convoluted nature of the rule-exception framework, and wondering "how much is enough, and how much is too much", one could question whether the AI Act is excessively detailed, potentially leading to a regulatory framework that is overly complex and requiring substantial interpretive effort for

[13] See https://www.tbs-sct.gc.ca/pol/doc-eng.aspx?id=32592 (accessed May 12, 2024).

implementation. Consequently, there is a risk that such meticulousness may result in fragmentation and, ultimately, overall uncertainty regarding the act. Clarifying boundaries between so many different situations may consume significant time and produce inconsistent outcomes.

3 The Artificial Intelligence Act

3.1 The Risk-Based Approach in the AI Act

A relevant factor contributing to the complexity of the AI Act is the risk-based approach, distinguishing between AI systems posing unacceptable, high, limited, and low or minimal risk, thus tailoring «the type and content of such rules to the intensity and scope of the risks that AI systems can generate» (recital 26 of the AI Act). This regulatory choice is consistent with the indication from the mentioned White Paper (European Commission, 2020) and with a worldwide regulatory trend for AI legislation (Kaminsky, 2023a). However, there are concerns that it would not guarantee a high level of protection of fundamental rights, as risk regulation operates on the assumption that a technology will be adopted regardless of its potential harms (albeit with a view to eliminate and mitigate them as much as possible) (Kaminsky, 2023b). Indeed, in the AI Act the concept of risk has, in fact, a twofold application.

Firstly, with a top-down approach, the European legislator has indicated different risk categories according to an abstract lists of use cases the AI systems may be applied to.

The Act explicitly prohibits harmful AI practices that create an unacceptable risk, such as the use of manipulative subliminal techniques and the exploitation of vulnerable groups.[14]

Moreover, most of the Act is dedicated to the regulation of high-risk AI systems,[15] i.e. AI systems that may have adverse impacts on safety and fundamental rights and that would have to adhere to certain requirements (such as transparency, human oversight and the establishment of a risk assessment system) and be subject to a conformity assessment before entering the EU market. Notably, the Act also sets out obligations for general purpose AI models (e.g. generative AI), which may be integrated into AI systems.

Finally, certain AI systems would have to comply with light transparency requirements (which may be cumulated with the high-risk ones), whereas low or minimal risk AI systems would not have specific obligations but would be encouraged to adopt voluntary codes of conduct.[16]

---

[14] Chapter II.

[15] Chapter III.

[16] The AI Act also proposes measures to support innovation, including the establishment of regulatory sandboxes for testing AI systems and puts into place a governance structure to monitor and impose compliance.

Despite the division into different risk categories, the rules established for prohibited and high-risk practices may prove ineffective in practice, because the risk level (and compliance therewith) is mostly self-assessed by providers (authorities enforcing the regulation would only control *ex post* in case of incidents or due to *ex officio* audits) (Veale & Zuiderveen Borgesius, 2021; Almada & Petit, 2023).

Secondly, there is also a facet of risk grounded into the practical design, development and deployment of the AI application, as providers need to carry out a risk assessment for the specific high-risk AI system[17] and the deployer (when a public entity or carrying out a public service) must perform a fundamental rights impact assessment regarding the context the system is implemented into. Indeed, all requirements for high-risk AI systems (e.g. representativeness of training dataset) are targeted to the intended purpose of the system (and related risks thereof).

Therefore, the AI Act, while also applying the precautionary principle by banning certain unacceptable risks, by its very nature allows AI systems with residual risks (e.g. low or minimal risk AI systems) to enter the EU market. Indeed, risk regulation tends to balance society-wide trade-offs, analyzing the aggregate impact of benefits and risks, and the AI Act aims to prevent (most) risks by prohibiting certain practices and ensuring proper design, development and monitoring of the systems. However, this approach does not seem to be equally well equipped to safeguard individual interests and (fundamental) rights. For instance, even though providers need to implement a risk management system for high-risk AI systems, some risks may only be mitigated, if they are deemed acceptable (what risk for fundamental rights may be deemed acceptable?) by the provider.[18] Notably, the Regulation explicitly foresees a situation in which a high-risk AI system, albeit fully compliant with the regulation, still presents risks (Art. 82).

At a closer look, the implementation of the Regulation also follows a risk-based approach, in a balance between legal certainty for businesses and fundamental rights protection. Indeed, after the Commission published the proposal in April 2021, negotiations between EU lawmakers finalized the legislation in December 2023, including amendments to the Commission's proposal. A staggered application recognizes the varying degrees of effort required to meet the different aspects of the regulation as well as the urgency of a prompt intervention in certain areas. After the entry into force of the Regulation, the transitional period will last (Art. 113):

- 6 months as regards prohibited activities;
- 12 months as regards provisions on governance and general purpose AI;
- 24 months as regards obligations pertaining to Annex III high risk AI systems and sanctions;
- 36 months as regards obligations pertaining to Annex I high risk AI systems.

---

[17] Risk assessments are also required for general purpose AI models with systemic risks.

[18] «The risk management measures referred to in paragraph 2, point (d) shall be such that relevant residual risk associated with each hazard as well as the overall residual risk of the high-risk AI systems is judged to be acceptable», Art. 9 (4).

3.2 *The Scope of the Regulation*

At the entry into force, the proposed AI Act would affect providers and users of AI systems established within the EU, with regard to AI systems utilized within the EU. Furthermore, it will also apply to providers and users of AI systems located in a third country, where the output produced by those systems is used in the EU (Art. 2).

The wide reach of the regulation may trigger the so-called Brussels effect (Bradford, 2012; Almada & Radu, 2024), which refers to the phenomenon where extra-EU countries often adopt EU standards to ensure access to the European market. As a result, the AI Act could potentially drive international AI companies to align with European requirements, even in non-EU jurisdictions, thus globalizing its influence. At the same time, it has been argued that conformity to the AI Act beyond the EU market's sphere of direct influence may also generate the side-effect of replicating some of the regulation's negative traits (Almada & Radu, 2024).

The AI Act is a horizontal regulation; therefore, it applies (Art. 2) to all sectors which are not explicitly excluded, i.e. research, testing and development activities or AI systems developed or used exclusively for military purposes or by public authorities in a third country.[19]

The Regulation will apply to AI systems or models introduced in the EU market (i.e. placed on the market or put into service) or, under certain conditions and with a scattered application, placed on the market before the entry into force of the Regulation.[20] Thus, the threshold for the application of the Act is the placement on the market[21] or the putting into service[22] of the AI system or model.

[19] Article 2. Nor to international organizations using AI systems in the framework of international agreements for law enforcement and judicial cooperation, art. 2 (4). Please note that in the AI Act, as in most EU regulations, article 1 identifies the subject matter, while article 2 defines the scope of the Regulation and article 3 sets out relevant definitions.

[20] Art. 111. Particularly, compliance is required by:

- 31 December 2023 for AI systems components of the large-scale IT systems (included in Annex X) and placed on the market or put into service before 36 months;
- 24 months from the entry into force for high-risk systems which underwent significant changes in their design;
- 6 years from the entry into force for high-risk systems used by public authorities; and
- 36 months from the entry into force for general purpose AI models introduced before 12 months from the entry into force.

Article 5 is directly applicable after 6 months from the entry into force for all AI systems (placed into the EU market before after the entry into force).

[21] «Placing on the market' means the first making available of an AI system on the Union market», Art. 3, point 9.

[22] «Putting into service' means the supply of an AI system for first use directly to the user or for own use on the Union market for its intended purpose», Art. 3 point 11.

3.3 *The Definition of AI Systems*

However, what is exactly an AI system? The definition of AI is highly controversial in the legal realm and even in computer science. However, it is worth remarking that defining AI within the AI Act is a policy choice, as it identifies the object and the scope of the Regulation and guarantees legal certainty (which supports innovation). Both broad and narrow definitions have pros and cons. On the one hand, a narrow definition (e.g. limiting the application of the AI Act to machine learning) would target the most advanced techniques (which are arguably more problematic because of the so-called black box problem) and would impose the design and development obligations only to their providers, thus reducing negative effects on the broad AI and software market.

On the other hand, a broad definition would ensure that even simpler systems do not present risks to rights and safety. Indeed, for the impacted person, the output is equally obscure and racial and gender biases in software predates (Lowry & Macpherson, 1988) the diffusion of machine learning techniques.[23] At the same time, the definition should aim to be as technology neutral as possible, as to include future innovations in the field (Renda & Engler, 2023).

Article 3(1) of the AI Act states that

> 'AI system' is a machine-based system designed to operate with varying levels of autonomy and that may exhibit adaptiveness after deployment and that, for explicit or implicit objectives, infers, from the input it receives, how to generate outputs such as predictions, content, recommendations, or decisions that can influence physical or virtual environments.

This provision modifies the initial definition proposed by the Commission, as the Parliament preferred to align it with the OECD definition (OECD, 2024)[24] in order to utilize a commonly recognized definition[25] and increase the Brussels effect.

The use of the term infer is very significant as recital 12 explains that inference directly relates to the capability of obtaining the outputs described above and the techniques that enable it go beyond basic data processing,

> include[ing] machine learning approaches that learn from data how to achieve certain objectives; and logic- and knowledge-based approaches that infer from encoded knowledge

[23] The issue of automated discrimination is not new: for instance, in 1970s and 1980s a Medical School in London utilized a screening program for job application which utilized surnames and places of birth to infer the ethnicity and the gender of the applicants. It was found to discriminate against women and ethnic minorities (Lowry & Macpherson, 1988).

[24] As modified in November 2023 (OECD, 2024).

[25] «The notion of AI system in this Regulation should be clearly defined and closely aligned with the work of international organisations working on artificial intelligence to ensure legal certainty, facilitate international convergence and wide acceptance, while providing the flexibility to accommodate the rapid technological developments in this field» (recital 12).

or symbolic representation of the task to be solved. The capacity of an AI system to infer transcends basic data processing, enables learning, reasoning or modelling.[26]

Moreover, the term autonomy refers to AI systems' degree of independence and operation without human intervention, while adaptiveness refers to possible self-learning capabilities during use.

AI systems may operate independently or as part of a product. Therefore, they may be physically incorporated into the product (embedded) or not. Indeed, the AI system may be embedded, downloaded in the user's device or accessed online, thusly being provided as a product or as a service. In the AI Act (consistently with as a trend in EU regulation of software), AI systems are always regulated as product, allowing a single regulatory approach regardless of the deployment method (Mazzini & Scalzo, 2023). The following pages will show the importance of the product legislation framework for the regulation of high-risk systems.

It's also worth noting that the AI definition also includes general purpose AI (e.g. generative AI). However, the Act also provides some *ad hoc* rules, which will be discussed in more detail below.

3.4 Prohibited AI Practices

As mentioned, the AI Act divides AI systems into different risk categories. At the highest risk level, systems posing unacceptable risks are explicitly banned from the EU internal market. The Commission initially identified four prohibited practices and, after negotiations between the Council and the Parliament, other four were introduced. The prohibited practices are listed in article 5 (Chap. II) and include:

- Real-time remote biometric identification systems in publicly accessible spaces for law enforcement purposes, unless the use is strictly necessary to achieve one of predetermined objectives,[27] and priory authorized. It is worth remarking that other types of facial recognition systems used for other purposes and in other contexts (e.g. border control, public transport, and schools) could be permitted as a high-risk AI system, subject to compliance with high-risk requirements.
- Social scoring, i.e. «evaluation or classification of the trustworthiness of natural persons over a certain period of time based on their social behavior or known or predicted personal or personality characteristics» (Art. 5, para 1). This practice is prohibited if, as a consequence, natural persons or groups are treated in an unfa-

[26] Recital 12. The initial proposal by the Commission included not only machine learning and logic and knowledge-based approaches, but also statistical approaches, Bayesian estimation and optimization methods, Annex I, point (c) of the proposal.

[27] E.g. «the targeted search for specific victims of abduction, trafficking in human beings or sexual exploitation of human beings, as well as searching for missing persons, Art. 5 (1) (h) (i).

vorable or detrimental manner in social context unrelated to the one in which the data to calculate the score was obtained and/or the treatment was unjustified or disproportionate to the social behavior or its gravity. For instance, an AI system cannot identify at-risk children based on their parent's divorce proceedings or by the fact that they missed a doctor's appointment.
- Deployment of subliminal techniques generating a material distortion of the behavior of a person, which causes (or is likely to cause) physical or psychological harm to that person or others.
- Exploitation of vulnerabilities of a specific group of persons identified due to their age, physical or mental disability when it generates the material distortion of the behavior of a person belonging to the group causes or is likely to cause that person or another person physical or psychological harm.
- Biometric categorization systems that identify sensitive traits (such as gender, race, ethnicity, citizenship status, religion, and political beliefs).
- Predictive policing tools based on personal traits or characteristics.
- Emotion detection systems employed in workplaces and educational settings.
- AI technologies that gather biometric data from social media or CCTV to form facial recognition databases.

3.5 *High-Risk AI Systems*

The complex regulation of high-risk AI systems is arguably the topic the AI Act poses most attention to. They follow the regulatory template of the New Legislative Framework for product safety. Therefore, high-risk AI systems can enter the market or be put into service only after they are compliant with certain requirements[28] and they underwent a conformity assessment. In particular, providers evaluate the conformity through internal controls, but they may be required to rely on external audits. Afterwards, providers affix the CE marking and the declaration of conformity and register the AI system in the AI Database. Then, the AI system can enter the market.

Furthermore, providers must ensure that AI systems perform consistently for their intended purpose and are in line with the requirements throughout their lifetime, evaluate continuous compliance with the Regulation and report serious incidents (post-market monitoring). The following figure represents the process providers should align to (Fig. 1).

AI systems are identified as high risk when they are safety components of a regulated product or they are themselves the regulated product, which may be subject to third-party assessment under the relevant sectorial legislation listed in Annex I. The

[28] Set in Sect. 2 Chap. III.

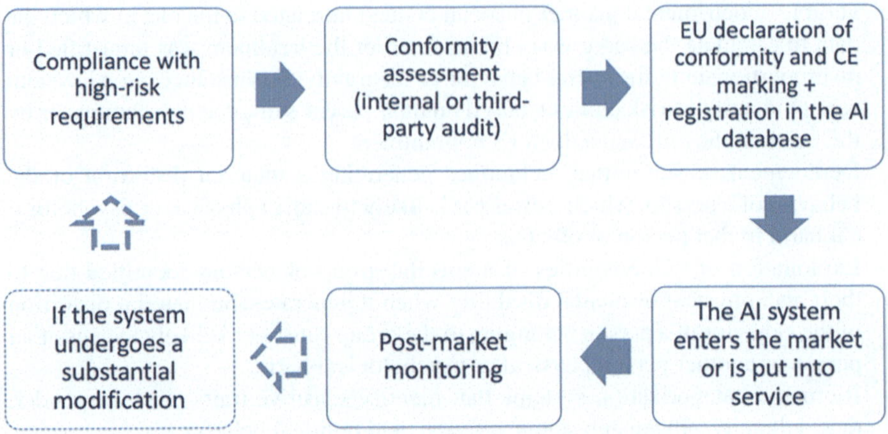

Fig. 1 Legal pipeline for providers of high-risk AI systems. Own elaboration

Annex comprises of product legislation in relation to, e.g. toys, medical devices, machinery and different means of transportation.

AI systems can also be considered high-risk when implemented in specific use cases included within the areas listed in Annex III (which may be extended by the Commission, pursuant to article 7). Therefore, when dealing with an AI system operating in one of the areas detailed below, it is necessary to check Annex III in order to see if the specific use case falls under the high-risk category.

For instance, an AI system assessing students in educational institutions is included in Annex III as a specific use case in the education and vocational training area, whereas an AI system intended to be used by authorities as polygraphs and similar tools or for emotion detection is included in Annex III as a specific use case in both the law enforcement and the migration, asylum and border control management areas.

The final text of the AI Act has included a further filter for Annex III use cases. Namely, AI systems may not be considered high-risk if they do not pose a significant risk of harm, to the health, safety or fundamental rights, «including by not materially influencing the outcome of decision making».[29] In particular, AI systems which perform narrow procedural tasks, improve the result of previous human activities, do not influence human decisions or do purely preparatory tasks are not considered high-risk.

However, AI systems performing profiling activities on natural persons are always considered high-risk when implemented in one of the use cases of Annex III.

[29] Art. 6(3).

3.6 *Requirements for High-Risk AI Systems*

As discussed above, high-risk AI systems will be permitted in the EU market only after demonstrating compliance with the requirements described in Sect. 2 Chap. III, i.e.:

- establishing a risk management system (article 9);
- ensuring proper data governance (article 10);
- drawing and maintaining technical documentation (article 11);
- keeping a record of logs (article 12);
- ensuring transparency and provision of information to users (article 13);
- setting up human oversight measures (article 14); and
- guaranteeing accuracy, robustness and cybersecurity (article 15).

Such requirements must necessarily be put into place before the AI system enters the market or is put into service. Some pertain directly to the design and development of the high-risk AI system, others establish documentation obligations in order to control and monitor the conformity of the system throughout its life cycle.

In particular, article 10 on data governance, establishes quality criteria (e.g. representativeness) for training, validation and testing of data sets and establishes that all high-risk systems must be subject to appropriate data governance and management practices.[30] Undoubtedly data governance is a crucial element against the replication and propagation of (both historical and possibly new, e.g. intersectional) biases and prevent AI discrimination.

Moreover, article 13 establishes that the design and development of high-risk AI systems should ensure that their operation is «sufficiently transparent» as to allow users to interpret and use the output of the system in an appropriate[31] way. Furthermore, instructions for use (in an appropriate format, digital or otherwise) must contain «concise, complete, correct and clear information that is relevant, accessible and comprehensible», including: the characteristics, capabilities and limitations of performance of the high-risk AI system,[32] predetermined changes to the AI system, the human oversight measures and the expected lifetime of the high-

[30] Such practices regard «(a) the relevant design choices; (b) data collection; (c) data preparation processing operations, such as annotation, labelling, cleaning, enrichment and aggregation; (d) the formulation of relevant assumptions, notably with respect to the informationInformation that the data are supposed to measure and represent; (e) a prior assessment of the availability, quantity and suitability of the data sets that are needed; (f) examination in view of possible biases; (g) the identification of any possible data gaps or shortcomings, and how those gaps and shortcomings can be addressed», Art. 10(2).

[31] The «appropriate type and degree» of transparency should also help achieve the compliance obligations of deployers and providers, Art. 13(1).

[32] And, in particular, its intended purpose, the level of accuracy, robustness and cybersecurity, any known or foreseeable circumstance, which may lead to risks to the health and safety or fundamental rights; its performance as regards the persons or groups of persons on which the system is intended to be used; when appropriate, specifications for the input data, Art. 13(3).

risk AI system (including any necessary maintenance and care measures). It's very important to note that the instructions are provided to the entity which will utilize the AI system in its professional capacity, not to the natural person who will be affected by the output. Indeed, the information contained therein are included to allow those operating the system to use it appropriately and prevent misuses and thus, this article does not require complete transparency (Busuion et al., 2022; Metikoš & Ausloos, 2025 including, for instance, a right to an explanation.[33]

Consequently, during the negotiations the European Parliament strongly campaigned to provide a «right to explanation of individual decision-making», which was included in Art. 86 of the final text. Such right is thereby granted to any affected person who is subject to a decision taken on the basis of an Annex III high-risk AI system when the decision produces legal effects or similarly significantly affects him or her (this wording replicates almost verbatim the phrasing of Art. 22 GDPR) «in a way that they consider to adversely impact their health, safety and fundamental rights».

The provision of article 13 on transparency is deeply interconnected with article 14, pertaining to human oversight. Indeed, human oversight is arguably the most evident concretization of the human in the loop principle. However, it is worth remarking the known overreliance issues in human-machine interactions, better known as *effet moutonnier* or automation bias,[34] which may transform a human oversight requirement from a user empowerment tool to a way to relieve the provider of responsibilities (Garapon & Lassègue, 2018).

Accordingly, article 14 requires both the actual oversight from an (adequately trained) natural person and the possibility of intervention from said person.[35] In fact, the design and development of high-risk AI systems should allow (including with appropriate interfacing tools built in the system by the provider itself when feasible)[36] their oversight from natural persons. Secondly, natural persons in charge of oversight must be able to: (i) understand the capabilities and limitations of the AI system (including the automation bias) and monitor its operation; (ii) ignore, override or reverse the result of the high-risk AI system; (iii) correctly interpret the high-risk AI

[33] In this sense, it has been pointed out that the concept of algorithmic transparency is transforming from the activity of "unveiling" and rendering visible the decision-making process that leads to a certain output to a communicative activity that allows, rather than comprehension, a form of easy interpretation of the output obtained by the user (Busuion et al., 2022).

[34] The *effet moutonnier* (herd effect) is a phenomenon in human-machine interactions whereby humans tend to endorse an automated system's output, potentially overlooking possible mistakes due to a lack of proper consideration and critical scrutiny. In particular, the book *Justice Digitale. Révolution graphique et rupture anthropologique* (Garapon & Lassègue, 2018) argues, with reference to predictive justice, that the output of an AI system may end up acquiring normative value due to the *effet moutonnier*.

Article 14 explicitly refers to this phenomenon as 'automation bias', which is referred to in the text as the «possible tendency of automatically relying or over-relying on the output produced by a high-risk AI system».

[35] Confirmation of at least two natural persons is required for biometric identification systems.

[36] When not feasible, the provider should identify the oversight measures which should be implemented by the user.

system's output; (iv) intervene on the operation of the high-risk AI system or interrupt the system through a stop button or a similar procedure.

However, risks can arise from the way systems are both designed and implemented and «choices in the governance structures can be instrumental in mitigating risks to fundamental rights in concrete use-cases»[37]; thus, the Parliament proposed the introduction of the fundamental rights impact assessment for all high-risk AI systems deployers (i.e. entities utilizing the AI system in their professional activity).[38]

Nonetheless, in its final version, the scope of application of this additional requirement has been considerably reduced. It applies only to high-risk systems as defined in Annex III and with the exception of the management and operation of critical infrastructure (Art. 27), when used by public bodies or private entities which provide public services, as well as by deployers using an AI system to evaluate the creditworthiness of natural persons or to carry out a risk assessment (or pricing) for the provision of life and health insurance. It is worth noting that, while the obligations in Sect. 2 Chap. III must be fulfilled by the provider, the fundamental rights impact assessment must be carried out by the deployer of high-risk AI systems, as the entity with the most complete knowledge on the concrete application and impact of the system.

Indeed, the assessment regards the «system's impact in the specific context of use» and includes (i) a clear outline of the intended purpose (comprising of the geographical and temporal scope), (ii) impact on fundamental rights and compliance with Union and national laws, (iii) categories of possibly affected people, (iv) risks of harm to marginalized people or vulnerable groups and adverse impact on the environment, and (iv) a detailed mitigation plan and a governance system (including oversight, redress and complaint-handling).

The provider also must carry out the conformity assessment (Artt. 16 and 43), which is the procedure that verifies the compliance of a high-risk AI system with the requirements of Chap. III, Section 2.

Compliance is assessed by either the provider itself (through the internal control procedure pursuant to Annex VI) or by a third-party, i.e. a conformity assessment body[39] pursuant to Annex VII. It is worth noting that, even when only internal controls are required, providers will arguably *de facto* rely on external validation established by private entities, for instance by adhering to a certifications scheme or by conforming to harmonized standards. In fact, except for high-risk AI systems listed in point 1 of Annex III (i.e. biometric recognition), in most cases, if a third-party conformity assessment is required but the provider applied harmonized standards or common specifications, it can also opt for an internal control based on Annex VI.

Hence, harmonized standards hold a very important role in the AI Act, as high-risk AI systems complying with them are presumed to be in conformity with the requirements set out in Section 2 of Chapter III (to the extent the standards cover those

[37] Recital 58a (European Parliament, 2023).

[38] Art. 3(4).

[39] Section 4 of Chap. III regulates the third-party entities which can attest to the conformity of high-risk AI systems.

requirements). Moreover, in areas not covered by harmonized standards, the Commission could adopt common technical specifications. Therefore, providers could self-assess their compliance on the basis of harmonized standards created by organizations composed of mainly industrial representatives (Eliantonio & Cauffman, 2020).

However, the utilization of harmonized standards has been contested because of the democracy deficit in standard setting organizations. In fact, these organizations comprise of industrial actors and guarantee little participation to other stakeholders such as consumer organization.

3.7 Limited and Low or Minimal Risks

Aside from unacceptable and high-risk AI systems, those presenting a limited risk would be subject to a limited set of transparency obligations regarding the nature of the system or of the output thereof (article 50). Therefore, this obligation differs much in scope from the transparency obligation provided for article 13 for high-risk systems. In particular:

- natural persons interacting with AI systems intended to interact with natural persons (e.g., chatbots) must be informed (and the systems must be designed and developed therefor) that they are interacting with an AI system (unless it is obvious from the context of circumstances). AI systems to detect, prevent, investigate and prosecute criminal offenses are exempted;
- natural persons must be informed by the user that they are exposed to an emotion recognition or biometric categorization system. Biometric categorization systems used to detect, prevent, investigate and prosecute criminal offenses are exempted; and
- users of deep fakes must disclose the artificial nature of the content. Deep fakes are defined as an «AI system generating synthetic audio, image, video or text content.[40] The content should be labelled in a clearly visible way to inform of the inauthenticity.

AI systems presenting only low or minimal risk (for example, spam filters or videogames) could be developed and used in the EU without additional legal obligations.

However, the proposed AI Act encourages providers of not high-risk AI systems to voluntarily apply the mandatory requirements for high-risk AI systems by adopting codes of conducts, which would be created by providers or associations thereof or other organizations (which may include other stakeholders). The regulation of AI systems presenting only low or minimal risks closes the provisions regarding the categorization of AI systems.

---

[40] Art. 50(2).

3.8 *General-Purpose AI Models*

The Regulation also sets out obligations regarding certain models. As already briefly mentioned, the *ad hoc* regulation of general-purpose AI models (in Chap. V) is an *unicum* in the AI Act framework, which otherwise uniformly regulates all AI technology, comprised in a wide definition aiming at technology neutrality as much as possible.

Indeed, the Commission proposal (back in April 2021) did not regulate general-purpose AI systems, other than with a cursory reference to the output of generative models (i.e. that the deep fakes have to be disclosed as such under Art. 50). Subsequently, the Council (in December 2022) (Hacker et al., 2023) and, most of all, the Parliament (in June 2023, after much debate was aroused by Chat GPT and other generative models) proposed different regulatory approaches (Olivato, 2023). This issue has arguably been one of the most contentious themes in the Trilogue negotiations (along with the inclusion of certain prohibited practices).

The final text regulates general-purpose models, which are defined by Art. 3 (63) as

> an AI model, including when trained with a large amount of data using self-supervision at scale, that displays significant generality and is capable to competently perform a wide range of distinct tasks regardless of the way the model is placed on the market and that can be integrated into a variety of downstream systems or applications, except AI models that are used for research, development or prototyping activities before they are released on the market.

Therefore, the legislator chose to also regulate general-purpose AI models, which are designed to perform a wide range of tasks without being limited to one specific application and can be included in AI systems (which may fall under the high risk category of the Regulation) to perform tasks within a specific domain or for a particular purpose.

A general-purpose model is considered to pose systemic risks when it presents high impact capabilities, which are evaluated on the basis of technical indicators and benchmarks set and updated by delegated acts of the European Commission.

Nevertheless, a presumption of high capabilities is set in art 51(2) when «the cumulative amount of compute used for its training measured in floating point operations (FLOPs) is greater than 10^25».[41] The Commission may also designate *ex officio* a general purpose AI model with systemic risk on the basis of the criteria set in Annex XI. Requirements for general purpose AI models with systemic risks are cumulative with the requirements for general purpose AI models. The Commission publishes a list of general purpose AI models with systemic risk.

[41] Such a threshold has been questioned as probably only a few models (data on the computing power utilized for training is often not public) would cross it (Hacker, 2023).

Requirements for general-purpose AI models mainly pertains to documentation obligations[42] and the respect of copyright and data governance in training.

The former includes information on the training and testing process and the results of its evaluation and other elements set out in Annex XI. The documentation must be provided, at request, to competent authorities (by an authorized representative, whose obligations are set out in Art. 54) and to providers of AI systems who intend to integrate the model.

Moreover, providers must set up a policy to respect EU copyright law and publicly release a summary on the content used for training. In addition, providers of general-purpose AI models with systemic risk must perform model evaluation through state of the art benchmarks and tools, identify, assess and mitigate possible systemic risks as well as reporting to competent authorities any serious incident (along with relevant corrective measures) and ensuring an adequate level of cybersecurity.

Compliance with the requirements for models can be demonstrated through adherence to a code of practice (until the development of a harmonized standard, whose compliance grants a presumption of conformity). Alternatively, the provider can demonstrate to the Commission that it undertook other adequate compliance measures. Codes of practice are regulated by article 56.

3.9 Regulatory Sandboxes

Measures to support innovation (Chap. VI) include the establishment of AI regulatory sandboxes[43] for the development and testing of innovative AI systems. At least one sandbox should be established (also jointly with other Member States) at national level and operational within 24 months from the entry into force of the Regulation (Art.57).

The regulatory sandbox would create a controlled environment for the development, testing, and validation of innovative AI systems before they enter the market (Ranchordas, 2021; Penasa, 2023). Sandboxes are operated under the direct supervision of the competent authorities (which are competent for the subject-matter of the sandbox, including national data protection authorities when personal data is processed) in order to guide the conformity of innovative technologies to the AI Act (Penasa, 2023). The authorities maintain their supervisory and corrective powers

[42] AI models provided under free and open license are exempted from the documentation obligation. However, they must comply with it if they are indicated as models with systemic risks, *ex* Art. 54 (5).

[43] Defined as «a controlled framework set up by a competent authority which offers providers or prospective providers of AI systems the possibility to develop, train, validate and test, where appropriate in real-world conditions, an innovative AI system, pursuant to a sandbox plan for a limited time under regulatory supervision», Art. 3 (55).

and any risk identified during the development and testing of the AI systems should be mitigated. Otherwise, the operations will be suspended.

The regulatory sandboxes should coordinate their activities and share best practices and recommendations through an annual report to the Commission.

The Commission will adopt an implementing act to standardize the access and operation of AI sandboxes in the various Member States.[44] Furthermore, the AI Act also provides the possibility of supervised testing in real-world conditions (i.e. not in a closed and controlled environment),[45] subject to informed consent given by the subjects of the testing.[46]

Other measures to support innovation (Art. 62) are tailored to support small-scale providers and startups, such as establishing priority access to regulatory sandboxes for SMEs and start-ups and supporting SMEs and start-ups in the organization of awareness-raising activities and the establishment of a dedicated channel to offer guidance and answer queries on the application of the AI Act.

Providing for a robust monitoring and evaluation mechanism is crucial to ensure that the proposal will be effective in achieving its specific objectives. The governance and enforcement structure put into place by the AI Act operates at two levels, European (Chap. VII Sect. 1) and National (Chap. VII Sect. 2).

3.10 Governance and Enforcement

At European level, the AI Act establishes a system for registering stand-alone high-risk AI applications in a public EU-wide database (Chapter VIII). Moreover, the European Artificial Intelligence Board (Artt. 65 and 66) is composed of representatives from Member States and the AI Office and supervises the application and implementation of the Regulation, *inter alia*, coordinating the efforts and assisting competent authorities, collecting and sharing best practices, provide recommendations and opinions. The Commission and the Member States will also be assisted on matters regarding the application of the Regulation by the Advisory Forum (Art. 67), which will comprise of representatives from several EU agencies as well as a selection of stakeholders, with a balance of commercial and non-commercial interests.

The Scientific Panel of independent experts (Art. 68) will support the Commission and Member States with regard to the implementation and enforcement of general purpose AI regulation. The AI Office (Art. 64), established within the Commission, will enforce and supervise the obligations pertaining to general purpose AI models.

At the national level, the notifying authority authorizes notified bodies to carry out conformity assessments and controls their operations (Chap. III Sect. 4).

---

[44] Art. 58.

[45] Art. 60.

[46] Art. 61.

The market surveillance authority is the authority in charge of the post-market monitoring and (i) assesses the compliance with obligations and requirements for high-risk AI systems; (ii) has access to confidential information, including the source code of AI systems; (iii) and would be required to take corrective measures for non-compliant or potentially risky AI systems. The functions of these two authorities could be reunited in only one authority.

Administrative fines of varying scales will be imposed as sanctions for non-compliance with the AI Act[47] and in particular:

- up to 35 million euros or up to 7% worldwide annual turnover for the preceding financial year, whichever is higher, in case of the violation of article 5;
- up to 15 million euros or up to 3% worldwide annual turnover for the preceding financial year, whichever is higher, in case of the violation of other provisions; and
- up to 7.5 million euros or up to 1% worldwide annual turnover for the preceding, in case of supply of incorrect, incomplete or misleading information to authorities.

For each type of violation, small and medium-sized enterprises (SMEs) will be subject to the lesser of the two specified fines, while larger companies will face the greater of the two amounts.

The Commission may impose fines, 1 year after the entry into application of the relevant provisions, on providers of general purpose AI model up to 15 million euros or up to 3% worldwide annual turnover for the preceding financial year, whichever is higher.

4 Concluding Remarks

In general terms, the AI Act demonstrates several points of interest. For instance, it is the first comprehensive piece of legislation in which a legal system (the European one) establishes binding regulations for the AI area. Regarding the characteristics that the law dealing with such divisive, pervasive, and transformative issues should possess (the so-called 3 Ts), the AI Act stands out for encompassing all of them (a Tuned, Timely, Tailored law). However, one might question whether the presence of a great amount of technical details, leading to numerous exceptions accompanying each rule, is excessive and thereby compromises the clarity of the regulation. Indeed, its complexity can certainly be seen as a double-edged sword. On the one hand, it allows the regulation of a wide range of applications and technologies in a comprehensive framework. On the other hand, its intricacies may pose challenges for compliance, potentially hindering innovation (especially for smaller entities).

Furthermore, the very protection of safety, health and fundamental rights may be hampered if the AI Act were reduced in practice to a mere compliance checklist and

[47] Chapter XII.

companies were able to exploit its many exceptions to their advantage (also by leveraging the role of industry-led standard-setting organizations). And the omission of environmental protection from the fundamental rights impact assessment is not a favorable indication.

Nevertheless, the AI Act is a significant attempt to provide a framework for AI regulation, setting a precedent for other jurisdictions and only time will tell its long-term effectiveness and impact on both innovation and the protection of safety, health, and fundamental rights. Its success will only be measured by the actual implementation and enforcement of the Act and by the ability of future revisions and adaptations to keep up with the evolution of technology and society.

The AI Act is far from a toothless beast (the Regulation set up hefty fines along with an important post-market monitoring system), however will the barking dog actually bite when the Act will enter into force?

Acknowledgements Some of the results presented by Carlo Casonato are part of the activities of the MUR PNRR project FAIR—Future AI Research (PE00000013) funded by the Next Generation EU.

Giulia Olivato's contribution was realized with co-financing from the European Union—FSE REACT-EU, PON Ricerca e Innovazione 2014-2020.

The views and opinions expressed are solely those of the authors and do not necessarily reflect those of the European Union or the European Commission. Neither the European Union nor the European Commission can be held responsible for them.

References

Almada, M., & Petit, N. (2023). The EU AI Act: A medley of product safety and fundamental rights?. *Robert Schuman Centre for Advanced Studies Research Paper*, 2023/59.

Almada, M., & Radu A. (2024). The Brussels side-effect: How the AI act can reduce the global reach of EU policy. *German Law Journal*, First View. Retrieved March 6, 2024 from https://www.cambridge.org/core/journals/german-law-journal/article/brussels-sideeffect-how-the-ai-act-can-reduce-the-global-reach-of-eu-policy/032C72AEC537EBB6AE96C0FD90387E3E

Barrat, J. (2013). *Our final invention: Artificial intelligence and the end of the human era*. Thomas Dunne Books.

Bostrom, N. (2014). *Superintelligence: Paths, Dangers, Strategies*. Oxford University Press.

Bradford, A. (2012). The Brussels effect. *Northwestern University Law Review, 107*(1), 3.

Busuion, M., Curtin, D., & Almada, M. (2022). Reclaiming transparency: Contesting the logics of secrecy within the AI Act. *European Law Open, 2*(1), 3.

Casonato, C. (2023). *Biodiritto*. Giappichelli.

Casonato, C., & Marchetti, B. (2021). Prime osservazioni sulla proposta di regolamento dell'Unione Europea in materia di intelligenza artificiale. *Bio Law Journal, 3*, 415–437.

Council of the European Union. (2022). *Proposal for a Regulation of the European Parliament and of the Council laying down harmonised rules on artificial intelligence (Artificial Intelligence Act) and amending certain Union legislative acts–General approach*, Interinstitutional file 2021/0106 (COD), 14954/22.

Crawford, K. (2021). *Atlas of AI: Power, politics, and the planetary costs of artificial intelligence*. Yale University Press.

Eliantonio, M., & Cauffman, C. (Eds.). (2020). *The legitimacy of standardisation as a regulatory technique–A cross-disciplinary and multi-level analysis*. Edgar Online.

European Commission. (2020). *White Paper on Artificial Intelligence–A European approach to excellence and trust*, COM (2020) 65 final. Retrieved March 6, 2024 from https://commission.europa.eu/publications/white-paper-artificial-intelligence-european-approach-excellence-and-trust_en

European Parliament. (2023). Amendments adopted by the European Parliament on 14 June 2023 on the proposal for a regulation of the European Parliament and of the council on laying down harmonised rules on artificial intelligence (artificial intelligence act) and amending certain Union legislative acts, COM (2021)0206.

European Union. (2024). Regulation (EU) 2024/1689 of the European Parliament and of the council of 13 June 2024 laying down harmonised rules on artificial intelligence and amending regulations (EC) no 300/2008, (EU) no 167/2013, (EU) no 168/2013, (EU) 2018/858, (EU) 2018/1139 and (EU) 2019/2144 and directives 2014/90/EU, (EU) 2016/797 and (EU) 2020/182 (artificial intelligence act). Official Journal of the European Union, L series, 1-144, 12.7.2024.

Fjeld, J. et al. (2020). Principled artificial intelligence: Mapping consensus in ethical and rights-based approaches to principles for AI. *Berkman Klein Center Research Publication*, No. 2020-1. https://doi.org/10.2139/ssrn.3518482.

Floridi, L., et al. (2018). AI4People—An ethical framework for a good AI society: Opportunities, risks, principles, and recommendations. *Minds & Machines, 28*, 689–707. https://doi.org/10.1007/s11023-018-9482-5

Garapon, A., & Lassègue, J. (2018). *Justice digitale. Révolution graphique et rupture anthropologique*. Presse Universitaires de France (PUF).

Hacker, P. (2023). What's missing from the EU AI Act: addressing the four key challenges of large language models, *VerfBlog*, Retrieved May 12, 2024, from https://verfassungsblog.de/whats-missing-from-the-eu-ai-act/. https://doi.org/10.59704/3f4921d4a3fbeeee

Hacker, P., Engel, A., & Mauer, M. (2023). Regulating Chat GPT and other large generative AI models. In *Proceedings: 2023* ACM conference on fairness, *Accountability, and Transparency.*

Harari, Y. N. (2016). *Homo deus: A brief history of tomorrow*. Harvill Secker.

Kaminsky, M. (2023a). *The developing law of AI: A turn to risk regulation*, The Digital Social Contract: A Lawfare Paper Series. Retrieved February 29, 2024, from https://www.lawfaremedia.org/article/the-developing-law-of-ai-regulation-a-turn-to-risk-regulation

Kaminsky, M. (2023b). Regulating the risks of AI. *Boston University Law Review, 103*, 1347.

Lowry, S., & Macpherson, G. (1988). A blot on the profession. *British Medical Journal (Clinical Research Ed), 296*(6623), 657–658.

Mazzini, G., & Scalzo, S. (2023). The proposal for the artificial intelligence act: Considerations around some key concepts. In C. Camardi (Ed.), *La via europea per l'Intelligenza artificiale* (pp. 2–5). Wolters Kluwer – CEDAM.

Metikoš, L., & Ausloos, J. (2025). The Right to an Explanation in Practice- Insights from Case Law for the GDPR and the AI Act. *Law, Innovation and Technology*, forthcoming

OECD. (2024). Explanatory memorandum on the updated OECD definition of an AI system. *OECD artificial intelligence papers*, no. 8, 2024. OECD Publishing.

Olivato, G. (2023). Paving the path towards general purpose AI systems regulation in the AI Act: an analysis of the Parliament's and Council's proposals. Media *Laws*, 3/2023.

O'Neil, C. (2016). *Weapons of math destruction: How big data increases inequality and threatens democracy*. Crown Publishing Group.

Penasa, S. (2015). *La legge della scienza: nuovi paradigmi di disciplina dell'attività medico-scientifica. Uno studio comparato in materia di procreazione medicalmente assistita*. Editoriale Scientifica.

Penasa, S. (2023). Verso un diritto "technologically immersive": la sperimentazione normativa in prospettiva comparata. *DPCE Online, 57*, 671–696.

Pound, R. (1923). *Interpretations of legal history*. The Macmillan Company. Retrieved September 8, 2024, from http://www.minnesotalegalhistoryproject.org/assets/Pound,%20Interpretations%20of%20Legal%20History%20(1923).pdf

Ranchordas, S. (2021). *Experimental lawmaking in the EU: Regulatory sandboxes* (Vol. No. 12/2021). University of Groningen Faculty of Law Research Papers.

Renda, A., & Engler, A. (2023). What's in a name? Getting the definition of artificial intelligence right in the EU's AI Act. CEPS EXPLAINERS. Retrieved March 3, 2024, from https://www.ceps.eu/ceps-publications/whats-in-a-name/

Russell, S. (2019). *Human compatible: Artificial intelligence and the problem of control*. Viking.

Russell, S., & Norvig P. (2021). *Artificial intelligence: A modern approach*. Global edition, 4th ed.

Veale, M., & Zuiderveen Borgesius, F. (2021). Demystifying the draft EU artificial intelligence act. *Computer Law Review International, 22*(4), 97–112.

Zuboff, S. (2019). *The age of surveillance capitalism: The fight for a human future at the new frontier of power*. Public Affairs.

Living in Digital Ecosystems: Are We Aware of This?

Stefano Cozzini and Mariarita de Luca

Abstract In this paper we would like to discuss and evaluate the social and economic impact of contemporary digital ecosystems. We extend the definition of digital ecosystem following the idea of the circular economy based on enhanced and efficient use/reuse of resources and products. We consider an Extended or Circular Digital Ecosystem, a Digital Ecosystem in which the final customers can create value by sharing their data and digital contents through the digital ecosystem services themselves. In this way the data consumers also become data producers. This new concept of a digital ecosystem is closely integrated with the role and actions of the people involved in it, as both users, producers, and managers. A key point of our discussion is the new and very powerful AI algorithms, which are the current focus for research and beyond. We argue that the focus should not only on the algorithms themselves but also be on the data used to train and/or feed such algorithms. We exploit the point of view of data mangers and data curators to address this issue. We make a comparison between the management of scientific research data that is moving towards correct management following the FAIR principles and the data produced every day on the internet by sharing information on social media or any other platform. Since 2009, following the idea of open science, the research community started to focus on the quality of the shared research data. In 2016, the FAIR (Findable, Accessible, Interoperable and Reusable) principles gave the guidelines to share data that are machine interoperable, i.e. that can also be correctly used and interpreted by machines and algorithms. For example, generative AI based on Large Language Models (LLMs), such as the well-known Chat GPT, was trained on a large corpus of text (about 45 TB) to understand and generate text consistently, which consisted mainly of any type of text obtained from web crawling tools. The users of digital ecosystems should be aware that all available data are currently used to train AI algorithms and the results that these algorithms produce depend on the quality of the data used to train them. It is widely demonstrated that database

S. Cozzini · M. de Luca (✉)
Area Science Park, Trieste, Italy
e-mail: stefano.cozzini@areasciencepark.it; mariarita.deluca@areasciencepark.it

A. Fabris, S. Belardinelli (eds.), *Digital Environments and Human Relations*, Human Perspectives in Health Sciences and Technology 150,
https://doi.org/10.1007/978-3-031-76961-0_6

classification by human intervention is prone to biases and errors. Algorithms trained on those datasets will suffer from these biases, resulting in dangerous or inappropriate content. Collecting all these insights, we plead that it is crucial to consider the technical experience of data managers and curators in the wider context of data sharing and management in society, given the immense amount of data produced and available on the network. We also argue that FAIR principles for research data should also be considered in the wider societal context, to avoid potential dangers related to new developments in Artificial Intelligence, which, in our belief and experience, are not harmful in themselves. Our general goal is then to raise awareness on the users of digital ecosystems that the correct sharing of their information can improve the quality of the data available on the network and that the potential dangers of data misuse can be reduced.

Keywords Digital ecosystems · Social and economic impact · Generative AI · Bias · FAIR Principles

1 Introduction

1.1 Big Data and Science

The digital revolution we are all experiencing is having a profound and pervasive impact in all our lives. The ability to acquire, store, manipulate, and transmit vast amounts of data has indeed transformed our world in ways comparable to the introduction of the printing press. The rate of production nowadays is impressive, and the term "Big Data" has been coined. Such term indicates in broad sense, the unprecedented flux of data stream toward large data infrastructure and feed by many different sets of devices and actors, including billions of people living on our planet. Big Data is therefore the engine of this digital revolution. It is worth note that the production of data can be active (i.e: agents create data on purpose) and/or passive (agent create data as side-effect and/or without consciousness). Big Data are commonly described by the four Vs:

- the Volume of data that data infrastructure must ingest, process and disseminate;
- the Velocity of data streaming through digital system in real time;
- the Variety of data to indicate the complexity of datasets in different format and structure that make their management and analysis a rather complex challenge;
- the Veracity of data (referring to the uncertainty due to bias, noise or abnormality in data). This last V has been introduced later as a desirable, not intrinsic feature, of Big Data. It is however a fundamental aspect as we will explain in the rest of this paper.

There are also other V-words, as reported for instance in (Leonelli, 2020). The most important to be mentioned is Value of data that is generally considered not as a feature itself but can be defined in many different ways depending as much on the intended use of the data itself.

Profound implications of the digital revolution are initially seen in science being researchers amongst the first users of digital networks and digital infrastructure. "Big Data" concepts, even maybe not defined in the way, were already familiar more than 30 years ago: Higgs boson discovery plays as perfect example: High Energy scientific communities discovered the particle after a large-scale data analysis of large datasets on dedicated digital distributed infrastructure. This specific example is, nowadays, no longer an isolated event: the increasing capacity of the computational infrastructure and the increasing rate of data production available to researchers of all disciplines are transforming many areas of research across a wide range of fields: from natural science up to humanities, and social sciences. In some research communities this process is already active, in others there is the possibility that access and analysis of such a large amount of data can trigger it. A new discipline, Data Science, is part of this transformation. Data Science can be defined as an interdisciplinary field that uses processes, algorithms, and systems to extract knowledge and insights (Data Camp) from the vast amounts of data available to make better decisions in many scientific communities. This is the so-called data-intensive science, as described in the famous book "The Fourth Paradigm: Data Intensive Scientific Discovery" (Hey et al., 2009), where a collection of papers discussed and elaborate the vision for a new, fourth paradigm of discovery based on data-intensive science. At the time of the publishing, fifteen years ago, still a limited number of people understood the impact on science we are experiencing today thanks to the Big Data revolution. To sustain this situation, the book proposed a shift towards multidisciplinary research as a key aspect of this new paradigm. It acknowledges that the complexity of modern scientific problems often requires expertise in multiple fields. This is particularly true with the rise of data-intensive science, where the ability to manage and analyze large datasets is crucial. The book's prediction that scientists would need to develop Big Data computing skills or collaborate with computing experts has certainly become a reality in many fields. There are along this line other authors that propose not a data-driven, or data-intensive approach to science but rather a data-centric approach to research (Leonelli, 2016). In their vision this approach has three distinct features:

1. data themselves are a research outputs and not only by-product of research: this implies that are valuable by their own;
2. the process of sharing, data is crucial for unlocking insights and driving discovery in various fields: this because shared data can be accessed by diverse individuals and organizations, the potential for new interpretations and discoveries grows exponentially and,
3. the first step of any research process is consultation of data resources through digital infrastructures and can guide researchers identifying novel research lines, and stimulate collaboration within the scientific community.

The possibility of linking together different findings (interoperability of the data) may facilitate the repurposing for a variety of goals that may have been unforeseeable when the data were originally generated (Leonelli, 2020). It is crystal clear that even if the digital revolution is far to be completed, it has already produced deep changes non only within the research environment we examined above but also in economic and in social environments. Big Data and, more recently the AI algorithms, have been promoted and perceived in the last few years with a strong positive approach: the well know refrain was and still is "data is the new oil". Indeed, this approach boost Big Data and AI as a social force for a positive change that contains the promise for unlimited use of an infinite resource (data) different from other natural resources that can be exhausted (i.e. oil). However, like industrialization, and oil in particular, data resources are not infinite, and the social changes caused by such resource and associated technology (not only AI but also edge computing/5G cellular network end other) on one side generate growth but on the other increasing drawbacks have been identified. The first that comes to mind is the impact on the natural environment in term of emissions and use of natural resource, i.e. the carbon footprint of Big Data and AI. Other negative drawbacks, sometimes less considered, is the impact on personal environment in term of individual right (for instance on protection and privacy of sensible/health personal data) on the broad social environment in term of democracy and unbalance of power. The combined effect of such negative impact is somewhere referred as data pollution, a concept we will discuss in detail in Sect. 5.

1.2 A Circular Data Ecosystem for Data Management

Big Data and Data Science require ad-hoc infrastructure, both in term of hardware and software and a dedicate and sophisticated set of tools to be put into production. These elements, together with the people involved in differs roles and with diverse contributions, will define what we name a data ecosystem. We will here discuss and analyze our personal vision of data ecosystem as a more specific definition of the standard digital ecosystem. A digital ecosystem can be defined as "a complex network of interconnected digital tools, including technology resources and digital platforms, partnering companies, and their services that interact with each other to function as a single unit. This ecosystem provides a seamless and integrated real-time experience for customers", (Digital Ecosystem). Following previous definition, a digital ecosystem can be represented with a graph where the knots or vertices are represented by companies, organizations, and people interconnected with each other. The connections represent the possibility of sharing data among nodes. Traditionally a digital ecosystem follows a customer-centric approach where each part focuses on providing value to customers. Another possible definition of data ecosystem is reported in (S. Oliveira et al., 2019) where we can read the following:

> Data Ecosystems are socio-technical complex networks in which actors interact and collaborate with each other to find, archive, publish, consume, or reuse data as well as to foster innovation, create value, and support new businesses.

From our perspective, we further extend the concept of a digital ecosystem and a data ecosystem following the idea of the circular economy based on enhanced and efficient use/reuse of resources and products (Berndtsson, 2015). We consider here an Extended Data Ecosystem or a Circular Data Ecosystem, a Data Ecosystem in which data is collected, shared, and used in a way that maximizes its value and minimize waste and negative impacts. The key concepts here is reusability that should effectively reduce the waste of resource needed to reproduce and recreate datasets already available but impossible to be found and accessed and/or simply too complex to define data access patterns. The above sentence contains three of the four words that defines the well-known FAIR (Findable, Accessible Interoperable, Reusable) principles of modern data management. The principles will be discussed in detail in Sect. 2; we here simply remark that they are embedded in our definition of Circular Data Ecosystem. It is worth now analyze which kind of resource can be wasted in reproducing data. Taking into account the first element of the ecosystem, i.e. hardware and more generally the hardware infrastructure, we can consider here the waste of computing time, storage capacity and network transfer. The carbon footprint of such waste is undeniable even if it is very complex to carefully evaluate it. We report in the following two different estimates to give a rough overview of such impact. The first one is provided by the French think tank advocating a shift to a post-carbon economy, the Shift Project. In their own words:

> the digital sector makes up a significant proportion of global emissions (3.5% in 2019) and a worrying growth of its impacts (+6% /year). Should we fail to reinvent digital behaviors and business models, the deployment of next-generation cellular networks – starting with 5G – and the subsequent development of uses (e.g., IoT, AI, edge computing, high-resolution mobile video feeds) are bound to aggravate this trend.

The second projection is taken from the International Energy Agency (IEA), an autonomous intergovernmental organization. In The Executive summary – Electricity 2024 – Analysis – IEA reports the following forecast for the next few years with respect to data centers (i.e. the global hardware infrastructure):

> Electricity consumption from data centres, artificial intelligence (AI) and the cryptocurrency sector could double by 2026. Data centres are significant drivers of growth in electricity demand in many regions. After globally consuming an estimated 460 terawatt-hours (TWh) in 2022, data centres' total electricity consumption could reach more than 1 000 TWh in 2026. This demand is roughly equivalent to the electricity consumption of Japan. Updated regulations and technological improvements, including on efficiency, will be crucial to moderate the surge in energy consumption from data centres.

2 Scientific Data Management and FAIR Principles

In the beginning of 2000, when the first Open Access Journals for result publications appear, the research communities, or part of it, began to wonder if and how to also make the research data accessible. In the paper "Data's shameful neglect" (Campbell, 2009) the benefits of data sharing are considered, especially when data are combined from many laboratories and analysed in ways that the original researchers could not have anticipated. More than 15 years ago, it was already clear that there was a need for specific instructions for sharing and managing research data, the need for competent people who could support researchers (data producers) in doing so and institutions or funding agencies to invest money to create the suitable digital environment. In 2010 a first attempt was the definition of "research objects" as a "semantically rich aggregations of resources, that can possess some scientific intent or support some research objective" and a list of principles and features to support not only the digital exchange of data, but the sharing of the digital knowledge related to those data (Bechhofer et al., 2010). The "FAIR Guiding Principles" were published in 2016, as a set of principles applicable to a "digital object" to make it Findable, Accessible, Interoperable and Reusable (Wilkinson et al., 2016). The "digital" attribute extends the previous definition of research object specifying that the object should possess a unique digital identity, i.e. it may have a machine actionable persistent identifier associated with it.

The great innovation of the FAIR principles was their specific emphasis on enhancing the ability of machines to automatically find and use the data, in addition to supporting its reuse by individuals, distinct from previous peer initiatives focused on the human scholar. The FAIR principles define the good rules of research data management, emphasizing that "good data management is not a goal in itself, but rather is the key conduit leading to knowledge discovery and innovation, and to subsequent data and knowledge integration and reuse by the community after the data publication process" (Wilkinson et al., 2016). Great strides have been made in terms of digital infrastructures for sharing research data in the last decade, but much remains to be done. In the (near) future the outcomes from FAIR data management will be high quality digital research publications that facilitate the ongoing process of discovery, evaluation and, reuse in downstream studies. The FAIR principles are a set of 16 suggestions for the management of research data and their metadata, defined as the data about data, indispensable for the correct interpretation of the data themselves. As mentioned above, the great innovation of the FAIR principles is related to the fundamental requirement that they are linked to both machines-driven and humans-driven activities. When dealing with digital objects, humans and machine face difficulties of different nature: humans have a strong semantic ability to recognize the appropriate context and correctly interpret even non-narrative information in the search for digital objects, while their limitation is related to the possibility of scaling and speeding up the operations necessary for the processing of large datasets, which is now a prerogative of the Big Data Science. To overcome these barriers, we need machines to be autonomously able to find, access, recognize, and operate with data; to achieve such a degree of machine actionability very high levels of standardization are required.

In the European context, providing fair and trustworthy data sharing environment is at the core of the European Commission's policy agenda for the period 2019–2024 under the priority "A Europe fit for the digital age". In 2020, the European strategy for data was published to propose the creation of common European data spaces in all strategic societal sectors and domains of public interest. Quoting the report of the European Commission: The idea behind the vision of the Data spaces is to build a trustworthy and interoperable data sharing environments where data can flow within and across sectors, in full respect of European ambitions, rules and values. (Farrell et al., 2023). The EOSC (European Open Science Cloud), the Data Space for research data, will provide researchers, innovators, companies and citizens with a federated open and multi-disciplinary environment where they can publish, find and reuse data, tools and services for research, innovation and educational purposes. The goal of EOSC is to develop a virtual environment which will become a 'Web of FAIR Data and services' for science in Europe upon which a wide range of value-added services can be built. These range from visualisation and analytics to long-term information preservation or the monitoring of the uptake of Open Science practices. This is an ambitious task that is still far to be completed and requires, in our opinion, a change in the mindset of many actors involved in the data life-cycles. We will discuss in some more details this aspect later in the Sect. 4.

In the following, we would like to point out that only few scientific communities possess extensive experience in managing, sharing and reusing research data. Their digital assets are meticulously curated and deeply integrated into specialized repositories tailored to their needs, such as Genbank (Benson et al., 2018), Worldwide Protein Data Bank (wwPDB) (Berman et al., 2009), and UniProt in the life sciences (Consortium, 2015); Space Physics Data Facility (SPDF) and set of Identifications, Measurements and Bibliography for Astronomical Data (SIMBAD) in the space sciences (Wenger et al., 2000). In these fields, the complexity and interdisciplinarity of research has imposed the need for adequate infrastructures for sharing research data. Other scientific communities are far behind and are only now beginning to address the challenge of sharing FAIR data and very few innovative institutions are training competent staff to support researchers in the data management and curation process. Moreover, general purpose repositories, such as Zenodo are emerging to offer suitable sharing services for scientific fields that do not fit in the data models of the special purpose repositories.

To accelerate and support this process towards the sharing of all research data, funding bodies have begun to require a Data Management Plan for publicly funded research projects. The Data Management Plan is a document where research projects should state clearly the way data are managed within the project itself. In its simplicity (especially for small projects) this document forces researchers to take into account from the very beginning the role of data in each phase of their research activities. This action of planning increases the awareness toward correct data management procedures.

While in H2020 Framework Programme (2014–2020) the Research Data Management (RDM) was an option, with the beginning of Horizon Europe Framework Programme (2021–2027), the European Commission has introduced mandatory requirements for FAIR data management for awarded projects and committed itself on the dissemination of data management information, making available the data.eu portal to share official publications, training opportunities and to build a community of aware users and producers of scientific data.

Considering that the goal of FAIR research data management is the advancement of knowledge through the reuse and sharing of results, we can consider other fundamental research aspects that derive from FAIR data management and reproducibility of data, such as research integrity. In the paper "Reproducibility and Research Integrity" (Resnik & Shamoo, 2017) the authors address the problem of data reproducibility, raising the issue that very often the lack of reproducibility is not just a scientific issue; it is also an ethical one. When scientists cannot reproduce a research result, they may suspect data fabrication or falsification. In several well-known cases, reproducibility issues led to allegations of data fabrication or falsification. In "Data Sharing and Data Integrity" (Chapter of the book "Data, Methods and Theory in the Organizational Sciences (Murphy, 2022)), a transparent process of sharing research data, under the more general umbrella of Open Science, can be the solution to eliminate the problems of data falsification and scientific misconduct. Certainly, FAIR data management alone cannot be the sole solution to combat scientific fraud. However, ensuring that the data linked to a scientific publication is Findable, Accessible, Interoperable, and Reusable significantly diminishes the likelihood of erroneous scientific findings.

To achieve a fully FAIR level of data management, it is necessary to make appropriate investments both at national and EU level, in (Mons et al., 2020) Barend Mons claims that 5% of all research funds should support FAIR by design strategies to ensure that research data could be reusable. Many researchers might argue that it would be funds diverted from research, but the price of not treating data in a FAIR way can be even higher. In 2018 the European Commission published a report on the "Cost of not having FAIR data" (Commission et al., 2018), estimating that the annual cost of not having FAIR data has the order of magnitude of EUR 10.2 billion per year. They considered five indicators to quantify the total amount: the time spent, the license cost, the cost of storage, the research retraction cost and the double funding. If we focus on the last one, the cost of solely double funding because of non-FAIR research is EUR 25 million per year, that represents an important amount of public money invested to reproduce already funded research. Many of the world's hardest problems can be tackled only with data-intensive, computer-assisted research and huge sums of taxpayer funds go to waste because such data cannot be reused. From this perspective, the investments in FAIR data management offer a dual advantage: they prevent the waste of public funds on redundant research and provide access to properly processed data that can be utilized by artificial intelligence algorithms. In fact, FAIR data allow much more effective artificial intelligence and playing with the acronym, Barend Mons claims that FAIR can be interpreted as "Fully AI Ready" (Mons et al., 2020).

3 FAIR for AI

In the previous section we extensively discussed about the FAIR (Findable, Accessible, Interoperable and Reusable) principles, introduced in 2016, as guidelines for suitable data management. The principles may also apply to any kind of digital object and may be extended to include the proper management of software (Chue Hong et al., 2022), tools, algorithms, and workflows that produce data. In the paper "FAIR for AI" the authors propose that the principles are now being adapted in the context of AI models and datasets (Huerta et al., 2023). It is not straightforward to define what FAIR means for scientific datasets and AI models. Differently from AI models trained on standardized dataset, the AI models for scientific discovery must incorporate specific domain knowledge to produce valuable results (Duarte et al., 2023). At the same time, synergies among different disciplines are fundamental to avoid the lead of knowledge silos, which would prevent the deployment of interdisciplinary FAIR AI ready datasets. The potential of AI model it is not predictable and the potential to enhance scientific discoveries is unimaginable, even in fields hitherto considered purely theoretical (Davies et al., 2021). Ensuring that data and AI models are FAIR facilitates the correct understanding of their content and context, enabling more transparent provenance and reproducibility. With a growing emphasis on data and its potential for novel applications, the idea that research data holds significant intrinsic value becomes increasingly evident, as already mentioned in the presentation of the data-centric approach. Scientific communities are transitioning from viewing research results solely in terms of scholar articles to recognizing data as autonomous valuable outcome. This shift in perspective suggests that in the future, FAIR AI research will lead the way for the development of high-quality, trustworthy, and reproducible AI applications.

We would like to highlight that within the scientific community there is an ongoing debate about how to produce FAIR datasets for AI tools. Many of these tools are based on algorithms whose functioning and results are testes again specific data samples (a process called "training"). We argue that AI algorithms trained on FAIR data could interoperate with all the correct information needed for accurate data comprehension and would produce trusted results. Additionally, this process would help to identify and mitigate potential biases that might otherwise influence the outcomes generated by these algorithms. Our concern revolves around the absence of similar discussions concerning AI algorithms trained on potentially biased standardized datasets. These models are ubiquitous in our daily lives, deployed across various technology platforms and accessed by millions of users.

AI algorithms need a huge amount of data to be trained and the quality of the data used has repercussions on the functioning of the algorithms themselves. Generative AI based on LLM (Large Language Models), as the well-known Chat GPT has been trained on a large corpus of text (an estimated 45 terabytes) to understand and generate text consistently. The parameters of the LLM are the different features of the language model, such as the ability to generate different types of text, the ability to translate languages, and the ability to summarize text. The more parameters an LLM has, the more complex it can be, the more tasks it can perform and more precise output it may produce. Those models are rapidly improving. GPT-1 had 117 million parameters to work with, GPT-2 had 1.5 billion, and GPT-3 arrived in February of 2021 with 175 billion parameters. It is not confirmed that GTP-4 has 1.76 trillion parameters (GPT-4 – Wikipedia).

Here, to highlight the problem, we briefly discuss the Common Crawl case, which has already been extensively analysed (Baack and Insights, 2024). Common Crawl is a small nonprofit organization that created a huge dataset (more than nine petabytes) crawling the web, back to 2008, and made it freely available. Recently, Common Crawl dataset has become one of the most important sources of training for generative AI. This dataset is deliberately non curated to leave to the downstream research the freedom to filter it according to the specific requirements of usage. Common Crawl mission is not aligned with the trustworthiness of AI algorithm, its goal is to make freely available the amount of data that usually only big company have access to. The consequences of this choice could be of potential danger for the end-users of AI algorithm trained on such dataset, not only in terms of harmful output, but in terms of biases incorporated in the crawled data coming from prioritized web pages on domains with many likes, not including domains preferred by marginalized communities. The long-term suggestion of the authors is to train generative AI algorithms on datasets more curated than Common Crawl and trace their provenance in more transparent way.

Also, datasets that were initially classified by hand potentially include dangerous biases that can alter the results of the AI algorithms trained on them. In the "The Data-Production Dispositif" (Miceli & Posada, 2022), the authors investigate how data for machine learning is produced through labour outsourced to Latin America. The investigation analysed task instructions, interviews with data workers, managers, and requesters, as well as observations at crowd-sourcing platforms and a business process outsourcing company. Their findings have shown that requesters use task instructions to impose predefined forms of interpreting data, introducing potentially harmful biases in the results of AI algorithms trained on those datasets, since precarious working conditions and economic dependency alienate workers, making them obedient to instructions. Furthermore, discourses and social contexts materialize in artifacts, such as interfaces and performance metrics, limiting workers' agency and normalizing specific ways of interpreting data.

These examples aim to underscore potential issues associated with generating the extensive datasets needed to train AI algorithms. We argue that, in many instances, unexpected outcomes are not attributable to the inefficacy of the algorithm itself, but rather to biases or defects introduced by the initial dataset. The extension of FAIR principles for any standardized dataset is not straightforward, we claim that with a bigger effort to manage and curate datasets for AI training, the potential harmful of AI algorithms can be controlled. Greater transparency on the details of provenance is essential, to know on which datasets the algorithms were trained and how those datasets were obtained. Within the European agenda "A Europe fit for the digital age", at the beginning of 2024 the European Data Act entered into force and probably on May 2024 the regulation on AI (AI Act) will be definitively approved. The purpose of European regulations is to deploy EU potential to become the global leader in safe AI. The AI tools and services will be discriminated on a risk based evaluation and to boost competition will be allowed the "sandboxes", considered controlled environments in which innovation is promoted and facilitated development, training, testing and validation of AI systems for a limited period of time before they are placed in the market or commissioning. The regulations should provide a strong regulatory framework based on human rights and fundamental values, to develop an AI system that benefit people, businesses, and governments. In (Soete, 2023) the author argued that, "*from the perspective of progress in research, science, and technology, rather than regulatory "quicksand boxes" for AI, public authorities and regulators should open the "black box" of data used to train AI and support algorithms*". As Barend Mons has usefully highlighted, "*good practices in data stewardship and analysis could be more effective than regulation.*" We agree that more transparency, more details on provenance, more semantic information, in practice a FAIR management of the data could improve the quality and the trustworthiness of AI application results.

4 The Different Roles and Responsibilities in Scientific Data Management

Research mission is to find new discoveries and to disseminate them to stimulate new open questions and fuel new discoveries. It is straightforward to think about a circular mechanism that potentially would never stops. Since data is a product of research, it is immediate to extend this circular process to them as well. In fact, as it is shown in Fig. 1, in research data management, we consider the data life-cycle as a circular process composed by different phases: Data Planning, Data Collection, Data Process, Data Analysis, Data Storage, Data Sharing and Data Reuse. Each of these steps must be realized following FAIR principles to be able to effectively reuse and share the research data. For each step, there are several actions required by researchers, data curators, data engineering, and there are many tools to ease each activity. Despite the tools available, and even though there are some very active scientific communities, still most researchers are reluctant to share their data and the cycle is very difficult to fully realize. If we focus on the people involved in each step of the research data life-cycle, the data producers and re-users of the data are usually the researchers, who are aware of their role, in the other phases, we can find researchers or data curators, data engineering, data stewards, etc.

We would like to keep our focus on the people and their roles within the life of the data produced in the daily activities. We recall here the idea of Circular Data Ecosystem we defined in Subsect. 1.2. There are data producers, intended as people which acquire and/or create data, people involved in the actual process for making data available to users (data suppliers) and then people which curate and manage data appropriately (data curators/data managers). Finally, there are data users and

Fig. 1 Research Data Life-cycle. Available with a Creative Commons Attribution 4.0 International License CC-BY by RDMkit

data consumers. Since, as an example, AI algorithms are fed and trained on large corpus dataset obtained crawling the web, the final customers and/or consumers can create a value by sharing their data and digital contents through the digital ecosystem services themselves. Then, the data consumers also become data producers, see Fig. 2. This process differs from the research data life-cycle, as in this scenario, the end user does not have the purpose of sharing the data to generate new knowledge, but regardless of the specific purpose, sharing data can still create value. The presented concept of a circular digital ecosystem is closely integrated with the role and actions of the people involved in it, as both users, producers and managers. This holistic approach where potentially all roles (data manager/data curator, data creator) can be played by the same people at the same time should bring within the Circular Data Ecosystem an awareness by all players about the critical aspects in data ecosystem.

In this context, it's worth sharing our personal experience regarding the interaction among various roles. We, authors of this contribution, are data engineers and data curators, but at the same time we produce and consume scientific data. This dual role contributes to our awareness on the importance of these instances while also enabling us to avoid compartmentalizing responsibilities and actions into very specific areas. In term of awareness of the importance of data: as both producers and

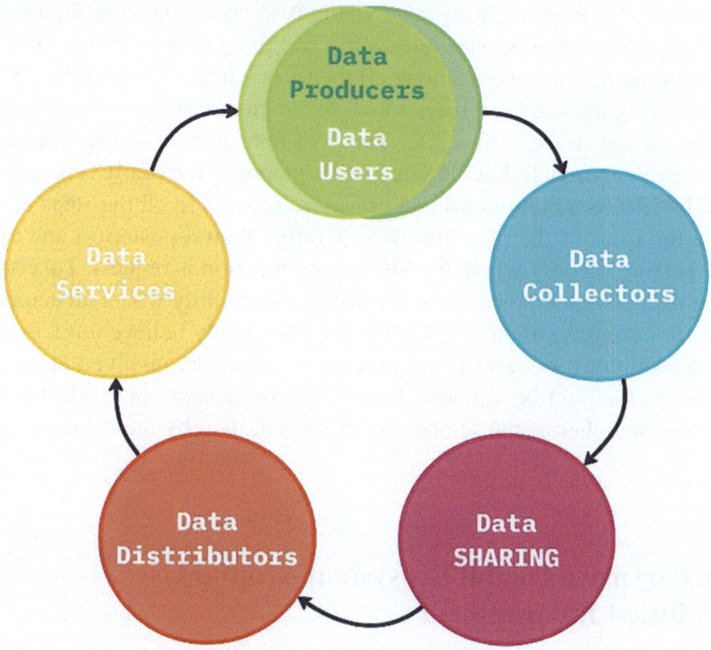

Fig. 2 Circular Data Ecosystem for data life and data roles for each action considered

consumers of data, we have a deeper understanding of the meaning of data in decision-making and value creation. This consciousness can lead to a greater focus on quality, integrity, and security of data throughout its life-cycle.

Another key concept to discuss here is the responsibility. This is especially related to our role of data-engineers, defined as the people who plan, develop, deploy, and maintain the hardware infrastructure and all the software layers above. We recognize the critical importance of our role in making informed decisions regarding infrastructure planning and ensuring the maintenance of the data ecosystem outlined above. Our ecosystem, even if limited to our labs and the infrastructure projects we are currently setting up, needs to be accurately planned to make sense of a messy sea of data. Research data are scattered in many different laboratories, highly fragmented, and poorly documented and curated. Moreover, there is a plethora of different formats and forms (images, video, text etc..) and there are large amounts of data coming from many disparate sources. The data ecosystem, we would like to provide, should consider all these aspects and difficulties to be compliant with the FAIR principles. Our final goal is to provide a data ecosystem where access and use stored data is no longer granted to small, privileged groups of scientists and researchers but to the whole scientific community behind the research infrastructure. This ecosystem can then facilitate the sharing of valuable and utilized data.

A key action for implementing such vision is to create the communities of both users and producers around specific topics and associated datasets. The process starts by stimulating awareness among all the roles within the scientific community and thus lowering the skeptical barrier that, based on our experiences, is still high in many actors active in data life-cycle. The goal here is to promote what we call a FAIR-by-design approach that integrate FAIR practices into daily research practice and make FAIRness a continuous and pervasive process in all the steps of research (and in all the roles of the data life-cycles), rather than a-posteriori and expensive operation performed only when data sharing/publication is request. This is possible only if our data ecosystem is built by the whole community where all actors play an active role in definition of such infrastructure. We firmly believe our Circular Data Ecosystem facilitates this awareness process of all actors involved and stimulates good practices that can be valuable for a FAIR-by-design approach that provides value to the researcher as the science is performed, thereby improving overall data quality.

5 The Pervasive Digital Ecosystem Around Us: A Polluted Environment

The Circular Data Ecosystem definition and considerations we elaborated in Sect. 1 for research data and research data life-cycle fit nicely also for the digital ecosystem that pervades any daily actions we performe for both working time and leisure times. Digital media are widely and heavily used and the three components (the data infrastructure, the software application and the people who produce, manage and use data) are also present. Hardware infrastructures in many cases are larger and more sophisticated than in research environment. The same is true for the software component which offers a wide scenario of sophisticated software tools. These tools are at same time extremely easy-to-use and engaging even for non skilled users. This combination in recent years has defined a data-driven technology landscape where data producers are digital citizens, many of them sometime and somehow not aware of this role. This is especially true for the so-called Data Exhaust. This term, as defined in (Schintler & McNeely, 2022) indicates a type of data that is often generated unintentionally by users from normal Internet interaction. It is generated in large quantities and appears in many forms, such as the results from web searches, cookies, and temporary files. Initially, data exhaust has limited, or no, direct value to the original data collector. However, when combined with other data for analysis, data exhaust can sometimes yield valuable insights. Such kind of data are therefore unwanted and many digital citizens are not fully aware of having generated them and having left a digital footprint that can be used by other (generally big companies) to generate valued services sometime paid by the very same citizen. As already stated in 2016 in "Data Ethics – The New Competitive Advantage" (Lapenta, 2016) data exhaust is not only undesired for the privacy and data protection rights of individuals, but also has an negative impact on the entire digital ecosystem of social institutions and public interest. This led Ben-Shahar to introduced data pollution concept in the legal field using again the refrain "data is the oil". In his own word:

> If indeed data are to this century what oil was to the last one, then I argue, data pollution is to our century what industrial pollution was to the last one. Pollution, whether industrial or digital, creates public harms and destroys public goods, in addition to the impact felt by private people who use products that pollute. The methods to control pollution and to protect public interests are distinctly different than the legal redress for private harm".

It is worth at this point introduce and elaborate on the concept of data pollution in details. Again, in Ben Share words:

> The concept of data pollution invites us to expand the focus and examine the ways that the collection of personal data affects institutions and groups of people beyond those whose data are taken, and apart from the harm to their privacy.

Figure 3, taken and slightly elaborated from (Hasselbalch, 2022) makes clear the three important environments affected by the data pollution problem: personal, social and natural. Eight different domains where data pollutant have important impact have been identified and discussed in the same reference. The catalogue comprises the following domain: nature, science&innovation, democracy, human rights, infrastructures, decision making, global opportunities, time. For each of them a short discussion is conducted on how data pollution can have adverse effects. As noted by authors such domains are described separately but they are often intrinsically interconnected. We review and comment briefly on such adverse effects for some of these domains, considering that they complement and integrate the discussion so far presented.

First, we consider here the human rights domain. Data pollution here refers to the fact that the well-defined human rights system established in real life is not yet fully implemented in the digital ecosystem we are part of. The increasing and the pervasiveness of digital technologies keep challenge individual human rights protection. The tension between the protection of individual human rights, such as the right to privacy, and the interests of National States and corporations to collect and control

Fig. 3 The data pollution schema with three environments affected

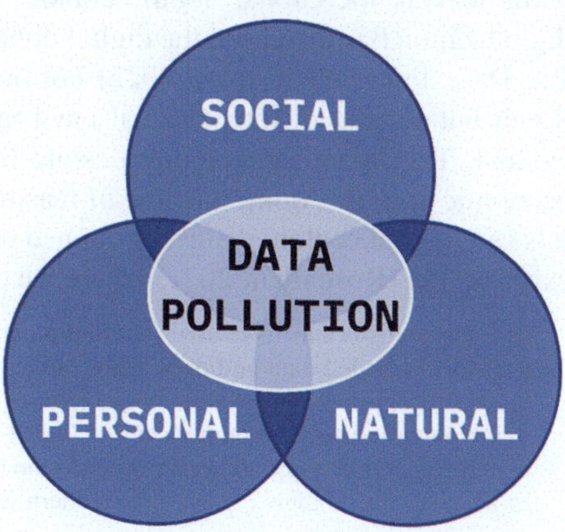

information is quite evident. States may argue that data collection is necessary for national security or to prevent crime, while companies may argue that data users are essential to provide personalized services or to their business model. This led to pernicious argument that privacy got a bad name for itself (Cohen, 2013).

From a European perspective there are concrete actions to mitigate such effect: the Charter of Fundamental Rights, approved in 2009 inserts the rights of EU citizens in EU law through article 8. The adoption in 2018 of the General Data Protection Regulation is another important step toward the human rights laws in Europe. At global level, it is worth mentioning that a few years before (2013), the United Nations General Assembly issued a resolution to affirm that the same rights that people have offline must also be protected online. This was necessary after the Edward Snowden's revelations of a mass surveillance intelligence global system.

It is vital therefore to have clear laws and regulations to protect human rights in the digital age. These laws should ensure that personal information is collected and used ethically and responsibly. In recent years, there has been a growing trend of challenging these issues affecting people's rights through legal instruments related to human rights in court. We mention here, as an example NOYB, a no profit company very active on this topic. It is worth remembering that its founder, Max Schrems, an Austrian lawyer sued Facebook Inc. to have illegally shared his personal data with the US. This had the notable effect to invalidate the Safe Harbour data-transfer agreement between the EU and US and led to the creation of the EU-US Privacy Shield.

As a personal reflection, this once again underscores the importance of awareness, every individual should be aware of their rights and how to protect their privacy online against the centralization and the intensive tracking of their own digital data. Here, we'd like to emphasize the significance of being aware of the impact of everyday digital action. We would like also briefly report on global opportunities domain; in this context, the author claims that data pollution has another effect: "it reinforces existing social hierarchies and colonial power dynamics that impact the distribution of global opportunities. The Big Data and AI 'revolution' has made the greatest difference in terms of opportunities in the economies of the Global North, while leaving the Global South behind.". We see here that unbalanced process of digitalization has increased the digital divide that is now expanded in what is named Big Data divide (McCarthy, 2022) not only among Global North and the Global South but also in different social environments where inequalities are wide and evident. This is still happening despite the promises of social opportunities and economic growth. Along this line of reasoning we would like to draw attention that this sort of "data colonialism", as defined in (Hasselbalch, 2022) reiterate a risk well expressed in (Rosenthal, 2021) where we can read the following:

> In a sense, historians have always relied on data exhaust to do our work, archives contain trails of discarded data and documents, albeit on paper. In my own research, I rely on the thousands of plantation account books left behind by American slaveholders, many of them highly quantitative. The numerical remains of this brutal economic system offer a cautionary tale for modern researchers. Scholars who have relied on slaveholders' data have tended to answer slaveholders' questions. Modern data sets are no less laden with values, and

revisiting the account books of American slavery can help us to recognize the risks of replicating the biases of our sources.

In closing this section we review again the infrastructure domain: the digital component may significantly alter our sociality with a digital extension of our real life and our social relations. Also the personal environment has been extended when identity can be revealed by profiling algorithm fed by exhausted data. This transformation relies on large data scale and computational infrastructures able to provide storage and computational resource needed to run AI algorithms and, constantly evolving and demanding more in term of natural resources need. Such infrastructures are dominated by a bunch of large companies and their carbon footprints.

Here, we'd like to emphasize the significance of being aware of the impact of our everyday digital actions. As already mentioned, this burden is increasing every day. Again, in this context, the role of data engineers and data scientist is important: there is a need of environmentally friendly data infrastructure developed by means of alternative technologies with a clear goal of reducing energy consumption. This can also be achieved by means of advanced energy-aware algorithm and systems. In this regard, data scientist could contributes developing alternative AI approaches that, for instance, do not rely on Big Data only. There are several innovative AI techniques and approaches (transfer learning, data labelling, artificial data generation and reinforcement learning) that can reduce training time and consequently the computational resources needed to train algorithms.

6 Conclusions

In this contribution we highlighted the importance of a FAIR approach to data management not only in a research environment, that sometime and somewhere is still missing, but also in our digital life. Annotating data with appropriate metadata and clear indication allows us, as scientists, to produce reproducible results. We argue that FAIR principles for research data should also be considered in the wider societal context, to avoid potential dangers not only related to new developments in Artificial Intelligence. We analyzed two different scenarios, related to data production:

- researchers produce a significant amount of data as part of their work, but their primary focus is often on the results of their research and the papers they publish. The data itself might not be their main concern, especially if they're not aware of the potential value or use of the data outside their specific research domain.
- people in their digital activities produce data unknowingly: this is a common occurrence in today's digital age as we discussed in previous sections. Most people are however not fully aware of the extent of the data they produce, nor do they consider its value.

In both scenarios, the value of the data itself might not be fully recognized or utilized. This is a significant issue in the field of Data Science and Big Data, where the potential insights from large datasets are often underutilized. It is of great importance for both individuals (as citizen but as scientist as well) and organizations to understand the value of data and consider the way they can be used responsibly and effectively. Our general claim is that raising awareness on the users of Circular Data Ecosystems about the correct sharing of their information could improve the quality of the data available on the network and reduce the potential dangers of data misuse. We argue here that awareness starts:

- when you understand that you are a data producer and not only a consumer: you leave behind you a path of exhausted data that are valuable for many actors;
- when you realize that data never comes for free. There is an associated cost to every digital action, and in a Circular Data Ecosystem sustainability must be considered.

Personal awareness and personal action can be taken by any digital citizen, but we also want to make clear that data ethics approach cannot be a single individual effort but need to be reinforced and completed by guidelines and rules in data management practices. These should become an important component in our Circular Data Ecosystem. Having an increasing awareness about the consequences and the cost of our interactions within our digital ecosystem we may mitigate the footprint of data pollution in all the tree environments mentioned in Sect. 5.

As already pointed out by many authors cited in this paper, an important reflection on the ethical and behavioral level is under discussion in the academic context. We claim here that it should be expanded at more general level to increase the chance of consciousness and awareness by any digital users. This is not only our personal vision. It is widely recognized, also in some more advanced and visionary companies, the urgency to implement an ethical and sustainable approach to data and AI. Nevertheless, despite of all the effort and global initiatives, highlighted above, we still need a much wider societal awareness to change our roles from data consumers to responsible digital citizens.

References

Baack, S., & Mozilla Insights. (2024). *Training data for the price of a sandwich.* Retrieved May 9, 2024, from https://assets.mofoprod.net/network/documents/Common_Crawl_Mozilla_Foundation_2024.pdf

Bechhofer, S., De Roure, D., Gamble, M., Goble, C., & Buchan, I. (2010). Research objects: Towards exchange and reuse of digital knowledge. *Nature Precedings.* https://doi.org/10.1038/npre.2010.4626.1

Benson, D. A., Cavanaugh, M., Clark, K., Karsch-Mizrachi, I., Ostell, J., Pruitt, K. D., & Sayers, E. W. (2018, January 4). Genbank. Nucleic acids research. *Nucleic Acids Research, 46*(D1), D41–D47. https://doi.org/10.1093/nar/gkx1094

Berman, H. M., Henrick, K., Nakamura, H., & Markley, J. L. (2009). The worldwide protein data bank. In J. Gu & P. E. Bourne (Eds.), *Structural bioinformatics* (pp. 293–303). Wiley.
Berndtsson, M. (2015). *Circular economy and sustainable development*. Retrieved May 9, 2024, from https://uu.diva-portal.org/smash/get/diva2:847025/FULLTEXT01.pdf
Campbell, P. (2009). Data's shameful neglect. *Nature, 461*(7261), 145.
Chue Hong, N. P., Katz, D. S., Barker, M., Lamprecht, A.-L., Martinez, C., Psomopoulos, F. E., Harrow, J., Castro, L. J., Gruenpeter, M., Martinez, P. A., Honeyman, T., Struck, A., Lee, A., Loewe, A., van Werkhoven, B., Jones, C., Garijo, D., Plomp, E., Genova, F., Shanahan, H., Leng, J., Hellstrom, M., Sandstrom, M., Sinha, M., Kuzak, M., Herterich, P., Zhang, Q., Islam, S., Sansone, S.-A., Pollard, T., Atmojo, U. D., Williams, A., Czerniak, A., Niehues, A., Fouilloux, A. C., Desinghu, B., Goble, C., Richard, C., Gray, C., Erdmann, C., Nust, D., Tartarini, D., Ranguelova, E., Anzt, H., Todorov, I., McNally, J., Moldon, J., Burnett, J., Garrido-S'anchez, J., Belhajjame, K., Sesink, L., Hwang, L., Tovani-Palone, M. R., Wilkinson, M. D., Servillat, M., Liffers, M., Fox, M., Miljkovi'c, N., Lynch, N., Martinez Lavanchy, P., Gesing, S., Stevens, S., Martinez Cuesta, S., Peroni, S., Soiland-Reyes, S., Bakker, T., Rabemanantsoa, T., Sochat, V., Yehudi, Y., & WG, R. F. (2022). *FAIR Principles for Research Software (FAIR4RS Principles).*
Cohen, J. E. (2013). What privacy is for (November 5, 2012). *Harvard Law Review, 126.*
Commission, E., for Research, D.-G., and Innovation. (2018). *Cost-benefit analysis for FAIR research data–Cost of not having FAIR research data*. Publications Office.
Consortium UniProt. (2015). Uniprot: A hub for protein information. *Nucleic Acids Research, 43*(D1), D204–D212.
Davies, A., Velickovi'c, P., Buesing, L., Blackwell, S., Zheng, D., Tomasev, N., Tanburn, R., Battaglia, P., Blundell, C., Juh'asz, A., et al. (2021). Advancing mathematics by guiding human intuition with ai. *Nature, 600*(7887), 70–74.
Duarte, J., Li, H., Roy, A., Zhu, R., Huerta, E., Diaz, D., Harris, P., Kansal, R., Katz, D. S., Kavoori, I. H., et al. (2023). Fair ai models in high energy physics. *Machine Learning: Science and Technology, 4*(4), 045062.
Farrell, E., Minghini, M., Kotsev, A., Soler Garrido, J., Tapsall, B., Micheli, M., Posada Sanchez, M., Signorelli, S., Tartaro, A., Bernal Cereceda, J., Vespe, M., Di Leo, M., Carballa Smichowski, B., Smith, R., Schade, S., Pogorzelska, K., Gabrielli, L. & De Marchi, D. (2023). European data spaces–Scientific Insights into data sharing and utilisation at scale, EUR 31499 EN. Publications Office of the European Union. ISBN 978-92-68-03166-7, https://doi.org/10.2760/301609, JRC129900.
Hasselbalch, G. (2022). Data pollution & power–White paper for a global sustainable development agenda on AI by Gry Hasselbalch with contributions from the Data Pollution & Power (DPP) Group at Bonn Sustainable AI lab.
Hey, T., Tansley, S., Tolle, K. M., et al. (Eds.). (2009). *The fourth paradigm: Data-intensive scientific discovery*. Microsoft Research Redmond.
Huerta, E., Blaiszik, B., Brinson, L. C., Bouchard, K. E., Diaz, D., Doglioni, C., Duarte, J. M., Emani, M., Foster, I., Fox, G., et al. (2023). Fair for ai: An interdisciplinary and international community building perspective. *Scientific Data, 10*(1), 487.
Lapenta, F. (2016). *Data ethics: The new competitive advantage*. PubliShare.
Leonelli, S. (2016). The philosophy of data. In L. Floridi (Ed.), *The Routledge handbook of philosophy of information* (pp. 191–202). Routledge.
Leonelli, S. (2020). Scientific research and big data. In E. N. Zalta (Ed.), *The stanford encyclopedia of philosophy* (Summer 2020 Edition). Retrieved September 5, 2024, from https://plato.stanford.edu/archives/sum2020/entries/science-big-data/
McCarthy, M. T. (2022). The big data divide and its consequences. In S. Carta (Ed.), *Machine learning and the City: Applications in architecture and Urban Design* (pp. 547–559). Wiley.
Miceli, M., & Posada, J. (2022). The data-production dispositif. *Proceedings of the ACM on Human- Computer Interaction, 6*(CSCW2), 1–37.

Mons, B., et al. (2020). Invest 5% of research funds in ensuring data are reusable. *Nature, 578*(7796), 491–491.

Murphy, K. R. (2022). *Data, methods and theory in the organizational sciences.* Routledge New York.

Oliveira, M. I. S., Barros Lima, G. D. F., & Farias L'oscio, B. (2019). Investigations into data ecosystems: A systematic mapping study. *Knowledge and Information Systems, 61*, 589–630.

Resnik, D. B., & Shamoo, A. E. (2017). Reproducibility and research integrity. *Accountability in Research, 24*(2), 116–123. PMID: 27820655.

Rosenthal, C. (2021, April 30). Lessons from the "data exhaust" of plantation slavery. *Harvard Data Science Review.* https://doi.org/10.1162/99608f92.c7418199

Schintler, L. A., & McNeely, C. L. (2022). *Encyclopedia of big data.* Springer.

Soete, L. (2023). Fair use in artificial intelligence? Access to data for the benefit of all. *Frontiers Policy Labs.* Retrieved September 5, 2024, from https://policylabs.frontiersin.org/content/commentary-fair-use-in-artificial-intelligence-access-to-data-for-the-benefit-of-all

Wenger, M., Ochsenbein, F., Egret, D., Dubois, P., Bonnarel, F., Borde, S., Genova, F., Jasniewicz, G., Laloe, S., Lesteven, S., et al. (2000). The simbad astronomical database-the cds reference database for astronomical objects. *Astronomy and Astrophysics Supplement Series, 143*(1), 9–22.

Wilkinson, M. D., Dumontier, M., Aalbersberg, I. J., Appleton, G., Axton, M., Baak, A., Blomberg, N., Boiten, J.-W., da Silva Santos, L. B., Bourne, P. E., et al. (2016). The fair guiding principles for scientific data management and stewardship. *Scientific Data, 3*(1), 1–9.

Navigating European Healthcare Regulations Through a Historical Lens

Leandro Pecchia and Alessia Maccaro

Abstract This chapter draws comparisons between significant EU legal frameworks, such as the Medical Device Regulations and the General Data Protection Regulation, and the socio-economic challenges that Europeans have faced historically. It uses these comparisons to shed light on the motivations behind the policymakers' decisions regarding upcoming regulations like the European Health Data Space and the AI Act. Additionally, the chapter outlines the similarities and differences between these regulations to assist scholars working with AI in healthcare in understanding the complex legal environment. The narrative begins with the establishment of the European Coal and Steel Community post-WWII, which was aimed at fostering Europe's physical reconstruction and maintaining peace; without coal, steel could not be produced, and without joint control over steel production, it was impossible for individual nations to manufacture war weaponry. In today's era, data is the new coal, and AI plays a vital role in driving European economic growth and competitiveness. It also holds the potential to become a formidable instrument of warfare. Therefore, ensuring equitable data sharing and ethical AI usage is essential for sustainable growth, peace, and democracy in Europe and beyond. The chapter concludes with a reflection on the inadequacy of existing regulations, particularly those in countries like the EU, USA, and China, to address low resource settings needs. It calls for future regulations to consider a more inclusive form of universalism.

Keywords European Health Data Space · AI Act · Regulations · Democracy · Universalism

---

L. Pecchia (✉)
Università Campus Bio-Medico Roma, Rome, Italy

Fondazione Policlinico Universitario Campus Bio-Medico, Rome, Italy
e-mail: leandro.pecchia@unicampus.it

A. Maccaro
School of Engineering, University of Warwick, Coventry, UK
e-mail: alessia.maccaro@warwick.ac.uk

A. Fabris, S. Belardinelli (eds.), *Digital Environments and Human Relations*,
Human Perspectives in Health Sciences and Technology 150,
https://doi.org/10.1007/978-3-031-76961-0_7

1 Introduction

The European regulatory framework for Medical Devices (MD) and In Vitro Diagnostics (IVD) has been going through consecutive changes since the 1970s, moving from a substantially national, subjective, and prescriptive approach towards a more international across all Member States and an objective approach, assigning more responsibilities to the manufacturer since the 1990s, and putting patients in the centre. This evolution was moved by the highest principles, ensuring the highest patient safety and defining clear responsibilities for manufacturers, but with the goal of building a unique EU marketplace and ensuring free circulation of MDs. Many authors refer to this shift using the expression "New Approach". According to the "New Approach," the responsibility of the manufacturer is balanced with the principle of free movement of products and the protection of health as a priority (Pallikarakis & Moore, 2007). The journey started with the Directive on active implantable MD (i.e., 90/385/EEC), followed by the Directive on MD (i.e., 93/42/EEC), and later the Directive on IVD (IVDs—98/79/EC). These Directives were supplemented by several Guidance documents on MD Classification, Vigilance and, in 2007, by a substantial amending Directive to resolve points related to the reclassification of implants, data confidentiality, clinical evaluation, post-market surveillance, and software validation (i.e., 2007/47/EC). Since 2007, medical software has been definitively equated to medical devices. In 2017, the EU MD Regulation (MDR 2017/745) and the IVD Regulation (IVDR 2017/746) replaced former directives, incorporating medical software into the definition of MD (Bini et al., 2023).

Scholars interested in impactful AI applications in medicine, and not only theoretical speculations, should familiarise with MDR, IVDR, but also with the General Data Protection Regulation (i.e., GDPR, EU Regulation 2016/679) as well as recent initiatives such as the European Health Data Space (EHDS) and the AI-Act, published in June 2024 (Fig. 1). EHDS and AI-Act perimeters have now been well defined and are clearly understandable in their principles and substance to those who are familiar with MDR, IVDR, and GDPR. In order to help scholars familiarise with this complex framework, the current paper offers historical prospectives, starting from the authors' assumption that the European Union was forged by the historical and economic challenges that Europe faced in the past 80 years, thereby

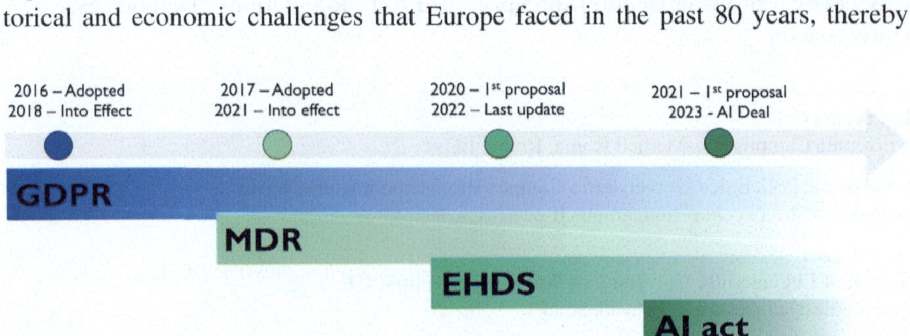

Fig. 1 Historical evolution of regulations

providing an understanding of the regulations in the context of this journey may help readers in facing the present and anticipating future regulatory changes.

2 The Journey from Steel to Medical Devices

2.1 Europe in the Post-WWII Reconstruction

After World War II, European Nations sought to rebuild their economies and establish stability. In 1951, with the Treaty of Paris, the European Coal and Steel Community (ECSC) was formed, aiming to integrate the coal and steel industries of six founding countries. This was a visionary and at the same time pragmatic idea: in order to reconstruct Europe after WWII, there was a significant need for coal and steel. Moreover, steel and coal were essential also to build weapons and move war, and with a shared management of those essential resources, no single country can make the weapons of war to turn against others, as in the past. Nowadays data are the new coal, and war leverage more and more on AI. Therefore, it is urgent that a new alliance on those two assets is defined in Europe and beyond.

2.2 Treaty of Rome and the Vision of a Common Market

The so called 'Swinging Sixties' was a period of unprecedented economic growth in Europe. In this context, the Treaty of Rome in 1957 established the European Economic Community, posing the bases for a common European market. This phase saw the elimination of trade barriers among member states and the gradual creation of a customs union. Economic integration deepened with the formation of the Common Agricultural Policy and the European Regional Development Fund, which were essential instrument in consideration of the economic and demographic shift that Europe was going through, including the baby boom, internal migration among wester EU countries, increased life expectation, urbanization. Yet, in the sixties, the Cold War exacerbated the economic divide among eastern and western EU countries, with a significant tension escalation since August 1961, when the first barriers of the Berlin Wall were built. Meanwhile, the EU expanded with the accession of new Western members, facing economic challenges such as oil crises and stagflation in the seventies. Efforts were made to address economic disparities among member states.

3 The End of Totalitarianism in Europe and the Need for New Movements for Goods and People

The eighties came with the first signs of the collapse of totalitarianism in Europe, since Solidarność trade union historical actions in Poland. There is nothing better than a common space for research, goods trade, and workers' free movement to consolidate Europe. On the 13th of June 1987, the Erasmus program was launched. This program has given so far to more than ten million people the chance to study, train, volunteer, and gain work experience abroad, significantly contributing to the making of a new European citizen generation. On the ninth of November 1989, when the Berlin Wall fell and with it, Germany was united after more than 40 years, and its Eastern half joined the European communities in October 1990. In the nineties, the former Yugoslavia began to break apart.

Now, there was a renewed need in Europe for a new model of freedom, allowing people, capital, goods, and ideas to move around European countries without barriers.

A major milestone was posed the seventh of February 1992, when the Maastricht Treaty was signed in the Netherlands, setting clear rules for the future single currency, foreign and security policy and close cooperation in justice and home affairs. The European Union was officially created, entering into force on the first of November 1993. The European Economic Area was created, entering into force on the first of January 1994 and since 1995, with the Schengen Agreement, in the first 7 countries (today 26), EU travellers were allowed to move with no passport controls at the frontiers. The euro was introduced in 1999 for electronic transactions, followed by physical banknotes and coins in 2002.

4 With Great Freedom Comes Great Responsibility: The Need for Common Rules on MD and Data

With the creation of a common space for citizens, industries, goods, and workers, goods produced in one country could be sold and used in another EU Country. This posed new challenges, especially in the family of products, bringing potential risks for end-users. This was the case with medical devices, where a significant knowledge asymmetry created a significant divide among designers, manufacturers, lead users, and end-users.

A renewed need for specific rules for the MD common market was now necessary. Therefore, few months after the enforcement of the Maastricht Treaty, EU rules on MD were harmonized with the three Directives regulating the common market and use of active implantable MD (i.e., 90/385/EEC), MD (i.e., 93/42/EEC) and in vitro diagnostics (IVDs—98/79/EC), replaced by MDR 2017/745 and the IVDR 2017/746 in 2017. Conversely, national regulations on hospitals and other health settings are regulated by member states and not equally harmonized in

Europe, although Cross-Border Healthcare Directive (2011/24/EU) establishes patients' rights to receive healthcare services in another EU member state and seeks to facilitate cooperation and coordination between member states in the provision of healthcare. This finds legal basis in the Maastricht treaty (article 152), which recognizes the autonomy of national health systems and promotes dialogue between social partners, materializing through the so-called "Open Coordination method".

There is already a significant bulk of literature describing the innovative contributions of the MDR with respect to the former 3 MD regulations, which readers can deepen (Melvin, 2022). The 2017 EU MDR ensures the safety and efficacy of MDs pursuing the highest level of protection for patient health, safety, and rights. Its primary purpose is establishing a comprehensive regulatory framework for medical devices within the European Union. The MDR introduces a more stringent risk-based classification system for MDs, leading to increased scrutiny and regulation for higher-risk devices. There is a heightened focus on clinical evaluation and post-market surveillance, requiring manufacturers to provide more comprehensive and transparent clinical data throughout the device's lifecycle. The regulation mandates using a Unique Device Identification (UDI) system, enabling the traceability and identification of medical devices throughout the supply chain. Notified Bodies, responsible for assessing conformity with MDR for riskier MD, undergo more stringent evaluation and oversight, ensuring higher standards and consistency in device assessments. Those stringent requirements for the MD market pose challenges for manufacturers operating in a global market: meeting the MDR's criteria might not align with the regulations in other regions or countries, creating a disparity in compliance standards and significant problems for lower-income countries surrounding Europe and during emergencies (Pecchia et al., 2020). Moreover, MDR does not mention AI, despite the fact that it has the potential to significantly impact several aspects of the regulation of MDR. This may require a reconsideration, and few amendments to the MDR, when the AI-act will be enforced. Conversely, the U.S. Food and Drug Administration (FDA) published dedicated plans for AI in relation to Medical Devices (U.S. Food and Drug Administration, 2020 [2020]).

5 From Coil to Data: The New Value Generation in Medicine

When Europe started moving from an "industrial economy" to a so called "knowledge economy" (Powell & Snellman, 2004), raw data became the new coil, and in order to generate value, there was a clear need to transform raw data into intellectual capital, innovation, information, education, training, research and development of new technologies. Therefore, again, it is not surprising if principles for data protection were defined with the European Union's Data Protection Directive 95/46/EC in 1995 (2 years after the MD Regulation), which was then replaced in 2016 (1 year before MDR) with the General Data Protection Regulation (GDPR), which came into effect in 2018.

5.1 *Legal Based of Fair Free Movement of Data in Europe: GDPR*

Regulation 2016/679 on the Protection of natural persons with regard to the processing of personal data and on the free movement of such data, commonly called General Data Protection Regulation (GDPR), was enforced the 25th of May 2018. The GDPR defines individuals' fundamental rights in the digital age; the obligations of those processing data (i.e., collecting, storing, sharing, analyzing data, etc.); methods and costs for ensuring compliance and sanctions for those in breach of the rules (Hoofnagle et al., 2019). Even if the referral to AI is not explicit, the GDPR is relevant for AI, because it includes the roles for fair data management and explanatory automated decision making and because it establishes the need for data protection impact assessments. However, GDPR is not an AI regulation; in fact, it is not specific to AI and does not cover AI aspects, except the process of personal data that AI uses. It is worth noticing that although the ambition of this directive was to regulate the free movement of data, to the extent that this was clearly mentioned in the title, too, there is a common impression that this regulation is mainly intended to protect data and limit their exchange. This resulted in many challenges, creating the need for further regulations, specifically in the sharing of health data.

5.2 *The Infrastructure for Sharing Health Data: EHDS*

Now that AI is promising to revolutionize many sectors, including medicine, in parallel with the need for a shared market of coil and steel, Europe is in huge need of sharing health data in order to maintain competitiveness and, in some areas, the leadership in medical research and MD manufacturing sector. The COVID-19 pandemic contributed to creating awareness in this regard. In 2020, it emerged a strong discussion around the need for the European Health Data Space (EHDS), aimed at creating a unified and interoperable ecosystem for health data management and exchange to improve healthcare outcomes, foster medical research, and enhance healthcare delivery, facilitating the secure and standardized sharing of health data across EU member states.[1] A core element of the EHDS is the distinction between primary use (clinical care) and secondary use (research, policy-making). The proposal of primary use is based on the building of a voluntary infrastructure named MyHealth@EU, not touching the national rules but prevising a new common legislative framework, where it is expected to provide better access to and exchange of electronic health data, with the final goal of providing better health outcomes, better evidence and saved costs. For secondary use, it establishes common EU rules on permits and common safeguards and unlocks the health sector's data economy

[1] See https://www.europarl.europa.eu/RegData/etudes/STUD/2022/740054/IPOL_STU(2022)740054_EN.pdf (accessed May 12, 2024).

potential through evidence-based policy-making and regulatory activities (Stellmach et al., 2022).

A further key element of the EHDS is the proposal of an "opt-in/opt-out" mechanism for using data following the GDPR. In this system, for primary use, the "opt-in" approach is required: the patient can control and, therefore, explicitly consent/dissent to the use of his data. An opt-in system is often seen as more respectful of individual autonomy and privacy, but it may result in further complexity in large dataset creation as it requires active consent registration from each individual. Conversely, for the secondary use of data is prevised an "opt-out" approach, where the individual can withdraw his consent, beyond the primary purpose of healthcare delivery, e.g., for research purposes. An opt-out system may lead to larger datasets creation as it does not require active consent from each individual. However, it may raise concerns about privacy and the awareness of individuals regarding the use of their data. The specific implementation of these mechanisms within the EHDS would need to carefully consider the balance between enabling robust research and innovation and safeguarding individual rights and privacy, which is a reflection of a more complex balance among citizens' right: right to the privacy vs the right to have more effective medicine, as result of impactful research (Saelaert et al., 2023).

EHDS is a pillar for the European Health strategy aimed at overpassing the rules existing fragmentation for building an overarching infrastructure for better health data management which can help in facing healthcare emergencies as the Covid-19 one. However, even the EHDS presents its limitations, in particular for the protection of sensitive information across borders, interoperability and data standardization because varied systems and practices among healthcare providers may hinder the seamless exchange and use of health data and the difference in national legislation can impede the harmonization of health data governance.

5.3 *Regulating AI for Medicine: The AI-Act*

The above scenario is completed by the recent news regarding the fact that the AI Act, which is the first-ever comprehensive legal framework on Artificial Intelligence worldwide, has been published. It was proposed by the EU Commission in April 2021, published for feedback and in December 2023 it was reached a political agreement between the EU Parliament and the Council, which was published the 13th, June 2024.[2] The new regulation is expected to be enforced after 3 years by 2027, as per any other EU Regulation. Yet, 4 years in this domain can be too long, therefore, readers may like to familiarise with the key elements of this regulation. The main new elements related to medical applications, and present in the

[2] See https://eur-lex.europa.eu/eli/reg/2024/1689/oj (accessed August 31, 2024).

provisional agreement,[3] are three: AI definition, AI EU regulation scope and perimeter, and the adoption of a risk-based approach as per MDR.

Regarding the AI definition, the AI-act adopts a specific AI definition, distinguishing AI from simpler software systems. The scope and perimeter of the AI-act has also been defined by the political agreement reached in December 2023, which clearly clarified that this regulation does not apply to areas outside the scope of EU law and exclusively demanded to member states, and clarification that the AI-Act will not apply to systems used for the sole purpose of research and innovation, or for people using AI for non-professional reasons.

Finally, as per MDR, the AI-act adopts a risk-based approach for AI. The AI-act defines clearly that AI applications that will not be permitted in EU. This includes AI applications for social scoring, for face recognition in public spaces, and for people manipulations. Yet, there are high risk AI applications, such as AI applications for educations, employment, justice, immigration, which are permitted, pending a conformity assessment, in parallel with Class II and Class III medical devices. The path for the conformity assessment will have to be carefully defined, but an equivalent of Notifying Body will have to be involved, after an accreditation procedure. Moreover, the AI-act regulates limited-risk AI applications, such as chat bot, deep fakes, emotion recognition systems not entering in the definition of unacceptable or high risk. For those applications, transparency will be crucial, with manufacturers required to use clear labelling, or disclosure that content has been manipulated (Veale & Zuiderveen, 2021). In line with the GDPR which already requires controllers processing personal data to be transparent about use of profiling and automated decision-making. Finally, minimal-risk AI applications include AI use for spam filters or videogames. For those applications the adoption of a code of conduct will be required, requesting manufacturers to adopt a code of conduct and avoiding any complicated CAP (Conformity Assessment Procedure).

5.4 The Contribution of the Global Community of Biomedical Engineers and Clinical Engineering

The European community of biomedical and clinical engineering has been proactive contributing to the development of the legal framework depicted in this paper. Founded in 2023, the European Alliance of Medical and Biological Engineering and Science (EAMBES; https://eambes.org/), a non-for-profit NGO, federates 67 biomedical engineering (BME) scientific societies and research institutions from 31 European countries, representing the European ecosystem of clinical and biomedical engineering. EAMBES is a member of the International Federation for Medical

[3] See https://www.consilium.europa.eu/en/press/press-releases/2023/12/09/artificial-intelligence-act-council-and-parliament-strike-a-deal-on-the-first-worldwide-rules-for-ai/ (accessed May 12, 2024).

and Biological Engineering (IFMBE; https://ifmbe.org/), which affiliates 6 international (IEEE MBES, AAMI, ACCE, CAHTMA, CORAL and EAMBES) and 77 national scientific societies from 74 countries, representing the global ecosystem of clinical and biomedical engineering. In this role, EAMBES supports the European Parliament and Commission in shaping the regulatory framework impacting patient safety and medical devices. Founded in 1959 in the UNESCO building in Paris, the IFMBE is a non-for-profit NGO in official relations with the UN World Health Organization (WHO), working with several UN agencies (UNESCO, ILO, WHO). Since their foundation, EAMBES and IFMBE have helped European and international policymakers in shaping the regulatory frameworks pertaining medical devices and allied technologies. In particular, since 2015, EAMEBS has been continuously supporting the European Parliament and the European Commission in developing relevant reports, such as the report on the Economic and Social impact of BME (Published in the Union Journal Eur-Lex, 2015/C 291/07), regulations, such as the Medical Device Regulations (2017/745 and 2017/746), the EHDS and the AI-Act, and in the creation of the first European Parliament Interest Group on Biomedical Engineering[4] (Pecchia et al., 2019). While BME community maintains its fingerprint and roots deeply into research, teaching and innovation, the fast evolution of medical device field requested those organizations to assume a growing and proactive role in supporting policymakers for shaping regulations with a significant impact on patient health and well-being.

6 Challenges

Each regulation mentioned in this paper plays a crucial role in shaping the landscape of healthcare, data privacy, and technology in the EU, with their specific focus areas and regulatory approaches. However, all these regulations started aiming at facilitating the creation of a unique EU market, for competing with emerging economies and large international players, by standardizing European practices, ensuring fairness, safety, privacy, and ethical standards across member states.

In particular, both the GDPR and EHDS deal extensively with data exchange, protection and privacy right. The MDR and AI-Act, while not primarily focused on data privacy, also have to take these aspects into account, particularly when dealing with health data or AI systems that process personal data.

About the regulatory approach used for MDR, GDPR, EHDS and AI-Act, there are differences which must be taken into account. The MDR and AI-Act include an explicit classification system based on low/high risk and the EHDS add the difference among minimal/limited/high/unacceptable risk. This then cascade CAP, defining clear path for ensuring safety and efficacy of products and services used in Europe. Conversely, the GDPR only refers to "simple" and "complex" data, with

[4] See https://eambes.org/epig-activities/ (accessed May 12, 2024).

	PURPOSE	KEY FEATURES	CHALLENGES
General Data Protection Regulation (GDPR)	• Subject rights • Fair data sharing • Purpose limitation • Transparency • Accountability • Legal basis for processing	• Principles: Data processing principles • People: new operators and responsibilities • Culture: data breach notification • Method: data minimization; proactive prevention for risk minimization • Tools: informed consent • Interoperability: cooperation; cross-border data transfer	• Global applicability • Conservative interpretation • Compliance costs • Regulatory Complexity • No explicit risk-based approach • Controls and data retention • Impact on innovation in EU • Implementation heterogeneity
Medical Device Regulation (MDR)	• Enhance patient safety • MD definition • Transparent info to patients, HPC and authorities • Improving transparency and accountability of operators • Improve clinical evidence	• Risk based MD re-classification • Conformity Assessment Procedure (COP) • From safety to efficacy approach • Stricter Market Requirements • Post-market surveillance • New responsibility for economic operators • Transparency and Traceability (UDI)	• Backlog of MD CE-mark • Awareness of operators • Notified Bodies' number • Burden on manufacturers • UDI implementation • Impact on EU innovation • Impact on EU SME
European Health Data Space (EHDS)	• Foster EU exchange of healthcare data • Enhance EU medical research • Improve interoperability • Create innovative health services • Patients' empowerment • Data governance	• Interoperability frameworks • Digital infrastructure • European Health Data Governance • EU Common data space • Enforce GDPR and improve cybersecurity • Data network • Data authorities • Digital innovation hubs • Olp-in/opt-out for primary/secondary use	• Data Security concerns • Implementation of technical and Infrastructure Challenges • Data ownership resistance • Trust perception • Data quality and standardization • Legal bases for data sharing/use • Ethical concerns for some data (e.g., genetic)
European Artificial (AI) Intelligence Act	• Enhance user safety • AI definition • Transparent info to end-users and authorities • Improving transparency and accountability of operators • Improve evidence	• AI-specific definition, regulations and requirements • Risk-based AI classification • Conformity assessment procedures • Transparency and accountability measures	• Interpretation Complexity • Data Access and Sharing • Overregulation • International Competitiveness • Enforcement and Compliance

Fig. 2 Comparative analysis among mentioned regulations

the result that people often interpret it in a "conservative" way this regulation in relation to health data sharing. GDPR does not explicitly define a low-risk use case, with an agile CAP, generating huge heterogeneity in its interpretation among each member state, and withing each member state among regions and hospitals. Figure 2 presents a comparative analysis among mentioned Regulations.

The MDR published in 2017, incorporated the 2007 amendment regarding medical software. Yet, it presented many limitations in regard to medical software, which will be exacerbated when most of MD will incorporate some AI. MDR regulation will require significant amendments and agile procedures for CE marking will be required considering novelties introduced by EHDS and the EU AI Act. MDR did not address AI, although AI applications in medicine were already significant in 2017.

A common limitation to MDR, EHDS and AI-act is the lack of explicit attention to the environmental aspect and social inclusion. Training AI algorithms has a significant CO2 footprint (De Vries, 2023) and sharing and storing data is a quite energy-intensive industry. As per MDR, it is surprising that after 30 years from the adoption of the first MD directives, there is now classification for MD regarding their CO2 emission. We need to make extra effort for ensuring that MD, EHDS and AI are green as other goods and services accessible in EU, and their production is aligned with EU net-zero emissions targets.

Equally important is the principle of leaving no-anyone behind, strength also by the 2030 Sustainable Development Goals (SDG). Instead MDR and GDPR created many de-facto barriers. CE marking for MD as defined in MDR is often used as a requirement for accepting MD in non-EU markets, as it is the case of many countries in Africa. Yet, most of African hospitals do not meet the same minimum-criteria as European hospitals, which are assumed as granted for EU MD manufacturers, exposing local patients and health-care operators to unnecessary risks (Di Pietro et al., 2020).

The AI Act responds to the need of a unique regulatory framework, but this is quite European-centric, while AI is a global challenge (Mügge, 2024). EHDS cannot be a lever to increase the digital divide with Low Resource Settings (LRSs) surrounding Europe, such as African countries. The risk is to repeat the mistakes of the past where universal regulatory frameworks fail to consider local and country context and specificity and were suffered from some areas. EHDS should already consider the integration of high-quality data from Africa with cooperation instruments. European innovators should use AI to reduce the digital divide among Europe and LRSs but this regulation does not open in this sense.

The origin of Europe and its institutions can be traced due to progressive historical-economic challenges: the urgency of post-war reconstruction, for which coal and steel were necessary, and the need to end the warlike climate pushed the countries to unite, sharing the management and circulation of those valuable goods. Subsequent historical developments confirm the growing need to facilitate the mobilization of goods and people within the European context.

Dematerialization of goods, services and values increased the importance of data: once again, wars of dominance and oppression must be avoided, and Europe needs to reaffirm its central positioning.

However, from the 1990s to 2017, many things have changed, and European legal framework initially emerged for an economic purpose have been refined, while others still need to be reviewed. MD become increasingly digital and software-dependent, paving the way for the transformation that is about to happen with AI, but with a substantial difference: when MDR was reformulated in 2017, the state of the art described in ISO standard and UNI norms was already wide and strong. Now EU policymakers are regulating AI, when the state of the art for this transformative technology is not yet well defined and far from being established. It is therefore natural for Europe to equip itself with strong instruments for health data sharing in to maintain competitiveness in respect to large American corporation and emerging Asian economies, which have different systems for protecting human rights.

The AI Act follows the now established method of European directives, which is the risk-based approach and clear CAP path, providing clear examples of what can/cannot be done and what can only be done following a strong control by notified bodies, which has worked well for MDs. In contrast, GDPR does not have an explicit risk-based approach, does not define explicitly low-risk cases, and is prone to heterogeneous interpretations.

6.1 Looking Forward to a Universalistic Approach

This chapter, aimed at outlining a historical-comparative horizon of existing European regulations on MDs, would not be complete without looking beyond Europe. Indeed, it is undeniable that, as of today, over 85% of the global MD market is governed by middle and high-income countries (USA, Europe, China, Japan, and Canada) that define increasingly higher standards and minimum requirements for medical devices, although more than 75% of the world's population resides in low and middle-income countries.

Most low-resource countries, in fact, are subject to such regulations and use medical devices designed for high-income contexts, which therefore have a much lower survival guarantee in different environments, compromising not only the effectiveness and safety of the devices but also patient rights.

In various papers, the authors of this chapter have attempted to demonstrate the existence of such a gap (Pecchia et al., 2020; Polisena et al., 2018). However, given the regulatory innovation described, it now seems truly impossible for new regulations, such as those on AI, not only to fail to mention medical devices but also to not genuinely include the needs of non-European contexts.

Including needs other than European ones does not mean yielding to relativism and abandoning a universal perspective. Universalization has a positive aspect of avoiding inequalities and creating uniform standards equal for all, but on the other hand, it has negative consequences such as proposing an approach so general that it becomes generic, i.e., incapable of being specific or adequate for very different contexts.

Therefore, only an approach that elsewhere we have defined as hermeneutic, heuristic, and ethical (Maccaro et al., 2022) will allow an effective change in interpretive perspective that, starting from the ground up, can view universality as a progressive, historical, dynamic tension towards an ideal to be achieved through continuous action of negotiation and mediation between particularistic demands and universalistic needs. This perspective aims at advocating for the rights of the most vulnerable and disadvantaged contexts, starting from a philosophical approach that recognizes the importance of particularistic specificity, the ethical relational tension, and global policies that abandon dirigisme for a directional attitude capable of guiding individual countries to bring out their own needs. Even with regard to ongoing regulatory evolution, it seems that such a framework could allow for a bottom-up approach aware of multiple contextual needs that fosters a genuinely inclusive universalism (Piaggio et al., 2021). This framework has been brought to the attention of international policymakers (Maccaro et al., 2023) and we hope that it will soon find a pathway for implementation that also intersects with the regulation of future AI discoveries, especially if applied to health.

Acknowledgements The authors express their gratitude for his thorough cross-review to Dr. J. Lovecchio (School of Science and Engineering, Reykjavík University, 102 Reykjavík, Iceland—Institute of Biomedical and Neural Engineering, Reykjavik University, 102 Reykjavík, Iceland).

References

Bini, F., et al. (2023). Is medical device regulatory compliance growing as fast as extended reality to avoid misunderstandings in the future? *Health and Technology, 13*(5), 831–842.

De Vries, A. (2023). The growing energy footprint of artificial intelligence. *Joule, 7*(10), 2191–2194.

Di Pietro, L., Piaggio, D., Oronti, I., Maccaro, A., Houessouvo, R. C., Medenou, D., et al. (2020). A framework for assessing healthcare facilities in low-resource settings: Field studies in Benin and Uganda. *Journal of Medical and Biological Engineering, 40*, 526–534.

Hoofnagle, C. J., Van Der Sloot, B., & Borgesius, F. Z. (2019). The European Union general data protection regulation: What it is and what it means. *Information & Communications Technology Law, 28*(1), 65–98.

Maccaro, A., Piaggio, D., Leesurakarn, S., Husen, N., Sekalala, S., Rai, S., & Pecchia, L. (2022). On the universality of medical device regulations: The case of Benin. *BMC Health Services Research, 22*(1), 1031.

Maccaro, A., et al. (2023). Pandemic preparedness: A scoping review of best and worst practices from COVID-19. *Healthcare, 11*(18) MDPI.

Melvin, T. (2022). The European medical device regulation–What biomedical engineers need to know. *IEEE Journal of Translational Engineering in Health and Medicine, 10*, 1–5.

Mügge, D. (2024). Eu AI sovereignty: For whom, to what end, and to whose benefit? *Journal of European Public Policy, 31*, 1–26.

Pallikarakis, N., & Moore, R. (2007). Health technology in Europe-regulatory framework and industry perspectives of the « new approach ». *IEEE Engineering in Medicine and Biology Magazine, 26*(3), 14–17.

Pecchia, L., et al. (2019). Health technology assessment and biomedical engineering: Global trends, gaps and opportunities. *Medical Engineering & Physics, 72*, 19–26.

Pecchia, L., et al. (2020). The inadequacy of regulatory frameworks in time of crisis and in low-resource settings: Personal protective equipment and COVID-19. *Health and Technology, 10*(6), 1375–1383.

Piaggio, D., Castaldo, R., Cinelli, M., Cinelli, S., Maccaro, A., & Pecchia, L. (2021). A framework for designing medical devices resilient to low-resource settings. *Globalization and Health, 17*(1), 64.

Polisena, J., Castaldo, R., Ciani, O., Federici, C., Borsci, S., Ritrovato, M., et al. (2018). Health technology assessment methods guidelines for medical devices: How can we address the gaps? The international federation of medical and biological engineering perspective. *International Journal of Technology Assessment in Health Care, 34*(3), 276–289.

Powell, W. W., & Snellman, K. (2004). The knowledge economy. *Annual Review of Sociology, 30*, 199–220.

Saelaert, M., et al. (2023). Expanding citizen engagement in the secondary use of health data: An opportunity for national health data access bodies to realise the intentions of the European health data space. *Archives of Public Health, 81*(1), 168.

Stellmach, C., Muzoora, M. R., & Thun, S. (2022). Digitalization of health data: Interoperability of the proposed European health data space. *Studies in Health Technology and Informatics, 298*, 132–136.

U.S. Food and Drug Administration. (2020). Artificial intelligence and machine learning in software as a medical device. Content current as of January 28.

Veale, M., & Zuiderveen, B. F. (2021). Demystifying the draft EU artificial intelligence act—Analysing the good, the bad, and the unclear elements of the proposed approach. *Computer Law Review International, 22*(4), 97–112.

For a Prudential Governance of AI

Vittoradolfo Tambone and Gianmarco Di Palma

Abstract In recent years, political interest in Artificial Intelligence (AI) governance has surged as AI is recognized as a General-Purpose Technology, impacting 40% of the labor market. The United States, Singapore, and the United Kingdom are leading in AI adoption, while global initiatives, such as those by the United Nations and the European Union, focus on establishing ethical standards and regulatory frameworks. The United Nations aims to serve as a universal governance body for AI, leveraging its global legitimacy, while the European Union seeks to create ethical AI standards to build trust. China and India are also developing their own AI strategies. The rapid expansion of AI raises concerns around data management, health, security, and ethics, intersecting with issues of democracy and corporate influence. However, AI also holds potential for positive societal transformation if governed responsibly. Effective AI governance must align with broader ethical goals, prioritizing inclusivity, sustainability, and the common good. A proposed "meta-project" for AI governance emphasizes defining a collective vision for AI's future and ensuring its development aligns with this vision. This requires multidisciplinary collaboration and a deep understanding of AI's societal impacts. Key principles include prudence, a human-centered approach, ecological sustainability, and making AI concepts accessible through popular culture. In conclusion, inclusive, transparent, and responsible AI governance is essential. It should prioritize the common good, align with the United Nations' Sustainable Development Goals, and address ethical, social, and practical considerations to ensure AI serves humanity's best interests and supports sustainable development and peace.

Keywords Artificial intelligence · Meta-project · Prudential governance · Ethical goals · Guiding principles · Sustainable development

V. Tambone (✉) · G. Di Palma
Università Campus Bio-Medico di Roma, Rome, Italy
e-mail: v.tambone@unicampus.it; g.dipalma@policlinicocampus.it

A. Fabris, S. Belardinelli (eds.), *Digital Environments and Human Relations*, Human Perspectives in Health Sciences and Technology 150,
https://doi.org/10.1007/978-3-031-76961-0_8

1 Introduction

Recent years have seen a rise in political interest in AI governance at different levels, considering the global revolution its use is leading to: a true worldwide paradigm shift, also because it is a "General-Purpose Technology (GPT)." Customarily, the dynamics of GPT development follow three phases: "In the first phase, while the technology is still new and not widely used, the benefits to productivity are small (…). During the second phase, the technology is improved upon, the cost of implementation tends to fall and the technology becomes more widespread, eventually leading to significant productivity gains. Finally, in the third phase, the law of diminishing marginal returns takes effect. At this point, the pace of improvements and roll-outs slows, causing productivity gains to taper off" (Capital Economics, 2024).

New technologies are, indeed, changing the global economy, thus far affecting about 40% of the labor market (Georgieva, 2024). According to the "AI Economic Impact Index", the U.S. is the leader in this transition, followed by Singapore, which has also demonstrated great adaptability and ability of diffusion as it is the center of the international market. In third place is the UK, which produces a lot of innovation even with a low investment index, partly because it is home to a world leader in AI such as Google DeepMind, based in London. Next come China, perhaps because of a lower diffusion of new technologies. Europe, being stratified by nation, does not appear in its overall size (Cazzaniga et al., 2024).

According to the International Monetary Fund, this dynamic entails both risks and benefits depending on individual and geopolitical characteristics (AI Advisory Body, 2023) and justifies the great political interest in the governance of this new revolution, which has produced a large number of initiatives in recent years aimed at proposing regulatory norms regarding the development and application of AI at all levels. In particular, the United Nations has come to propose itself as an advocate for a Universal Government of AI for the good of Humanity, which seems to us a particularly interesting example. In December 2023, the Advisory Body formed by the UN for this purpose, presented an Interim Report where this proposal has become explicit, even if still offered as such to the international debate. Quoting the Document: "The United Nations lies at the heart of the rules-based international order. Its legitimacy comes from being a truly global forum, founded on international law, in the service of peace and security, human rights, and sustainable development. We believe that this offers the institutional and normative basis for collective action in global governance of AI" (European Commission, 2021a). For its part, the European Union has been working for several years on possible regulatory standards, aiming to characterize European AI as AI "Ethical" and, therefore, more capable of building trust in end-users. On April 2021, the European Commission produced the "Communication on Fostering a European approach to Artificial Intelligence" (European Commission, 2021a), the "Coordinated Plan on Artificial Intelligence" (European Commission, 2021b) and, on a more properly regulatory level, the "AI Act" (European Commission, 2024a). All this was followed by the strategic plan "Artificial Intelligence in the European Commission (AI@EC) Communication", which was proposed as a global regulatory plan (European Commission, 2024b) similarly to what the United Nations had previously done. At the beginning of February 2024, all member countries accepted the Draft of the heralded AI Act, which was published the 13th, June 2024 (European Union, 2024).

For its part, without undue fanfare, China as early as 2017 "announced an ambitious program for the domestic development of artificial intelligence (AI) technology, aiming to become the world's 'leading AI innovation center' by 2030. China plans to expand AI into many spheres of manufacturing, governance, and defense by that deadline" (European Parliament, 2021). On the other hand, the Indian government presented its strategy regarding the development of AI in 2023 emphasizing the landmark philosophy of "Sabka Sath, Sabka Vikas Aur Sabka Vishwas" i.e., of "together with all, development for all and trust of all" (Government of India, 2023).

At the same time, there have been international meetings in which government leaders and entrepreneurs involved in the development and commercialization of AI have also proposed recommendations for AI governance with significant media coverage (AI Safety Summit, 2023). Although many other national and international documents concerning AI governance could be mentioned, the foregoing seems to us sufficient to witness to the enormous global interest that exists to legally, politically and therefore economically control the development, management, and use of AI. When we speak of economic interests, we of course refer to all areas, from defense to health, having a significant influence on markets, at micro or macro level. Indeed, AI has a systemic impact that the European Union seems to have grasped when it talks about the Systemic Risk, especially related, to a certain type of generative AI.[1] All this suggests that the issue of the type of AI governance is one of the main ones to be currently addressed.

2 Prudential Governance of AI

Although the dystopian horizon of the "Technological Singularity" (Kurzweil, 2006), and thus of an auto-poietic technology, remains little more than literature, on the other hand, the effectiveness and efficiency of new technologies are already a fact of life at all levels: medicine, economics, communication, defense, etc.... Such a pervasiveness of artificial intelligence has primarily led to a huge increase in the amount of data available and the need to use it correctly and efficiently through their ever-improving analysis and interpretation (Ebbert-Streib, 2021), as well as to an ethical alert regarding the use of this information often concerning sensitive areas such as health and security (Kumar et al., 2020). Besides fostering the ubiquitous

[1] Generative AI at Systemic Risk is, for now, the one with a cumulative computing capacity used for training greater than 10^25 flops and other efficiency indicators (See Draft of AI Act for completeness).

application of AI, the Internet of Things further encourages the flow of data through the use of sensory data and the employment of ubiquitous devices producing zettabytes of data in real time (Baccour et al., 2021). This development also pervades and affects the political debate, especially with regard to the respect for constitutional democracies (Nemitz, 2018) and its oligarchic and corporate development (Baradaran, 2024). This trend already bears within itself the impetus for even more pervasive development, in view of the use of next-generation mobile networks for an even closer and more functional connection between the human, physical and digital dimensions (Filippou et al., 2022).

Facing the development of AI makes us feel like one of the little boats coping with the Great Kanagawa Wave (Hokusai, 1830–32). Faced with this real tsunami, the attitude may be the urgent desire to stop it[2] (the wave would sweep over the boat, tearing apart), or "free-diving" until it passes (hoping that it is not a too long wave to lead us die drowning). Still, trying to ride this wave by inventing a kind of "Surf"[3] seems to be the best attitude indeed! And so, thanks to our steering skills and a simple little boat, this way would become a new way to nearly fly and experience something previously unthinkable.

It is not about "putting on a good face" by pretending not to see existing dangers in the pathetic illusion that this is enough to exorcise them. Nor does it mean that we settle for regulating, as best as we can, the evil we may encounter in the world since it is impossible to cope with it, thus surrendering to an inevitable situation. It means, instead, starting from the real in order to experience it as a resource in view of a positive overall project, a meta-project, capable of directing it effectively and positively.

It is about developing the "political intelligence" which, for Aristotle, gives us the ability to deliberate wisely (Aristotle et al., 2009) what we desire righteously because it is good: "Now, in political deliberation the thing known at the outset, which serves as an indemonstrable principle, is the good end, that is, the kind of things that constitute the object of desire when it is righteous" (Bodeus, 1993). When we talk about a meta-project, we are referring to this "good end" that allows us to understand what means to use it and to improve it as is appropriate, through prudential judgment. This political intelligence engages the freedom of the human being (HB) with a creative dimension that, for Hannah Arendt, characterizes the best politics (Arendt, 2005), namely that which starts from the idea that "the meaning of politics is the freedom" (Arendt, 2005 [22]) of unique human beings who are not massified into an anonymous concept of "humanity".

---

[2] This degenerative approach is manifesting itself at various levels, emphasizing the need to build "guard-rails" for AI. Internationally, a moratorium has been proposed several times to slow down the development of new technologies. In line with Lakatos' thinking, it seems to us that projects aiming to impose brakes and define prohibitions are losers compared to proactive projects that, instead, shape reality in a positive way.

[3] As well as trying to surf (London, 1911).

Put another way, the interest we have in AI is not directly in it but in the idea of the World that we find see possible to realize even through AI, which, inversely, we intend to develop and govern in a way that is synergistic to this meta-project of ours.

In recent years, technological research has used a multidisciplinary methodology (pluridisciplinary, trans-disciplinary, interdisciplinary, or antidisciplinary, if you will) that has made computer science, mathematics, robotics, philosophy, and law dialogue with the world of experimental science. However, it has never been an area of cultural interest only. To put it mildly, Good Technology has never been interested in technology for its own sake, but it has always wanted to shape society according to a very specific vision of the World. In a sense, from the beginning technology has also been a political tool, meaning "politics" as the highest expression of Ethics (along the lines of Aristotle). Technology is not exclusively concerned with technology itself, but also with Economics, Medicine, Engineering, Defense, Philosophy, Science, Law, Politics, Art … and, in order to govern technology, one must do so with the awareness of the end one is aiming at and for which technology will be a useful tool. In other words, it is a matter of being aware of the unifying goal. This unifying element, the "good end" just mentioned, is what we mean by meta-project, seen as an overall socio-political goal to be achieved through appropriate specific objectives of the sciences involved and brought into synergy. This meta-project will give systemic, proactive effectiveness to the work done by each particular science (whether experimental or speculative), by entering into coherent dialogue with the different sciences involved with feedback that will appear as confirmations or corrections. Therefore, the worldview to be proposed will be multidimensional, scientific, cultural, artistic, political, economic, and theological. This convergence of opinions makes the proposal strongly credible. The unification of different scopes through a meta-project makes whatever is done of interest to the specific interlocutors. At the same time, this contributes (or it makes possible a future collaboration) to the strengthening of the joint endeavor for the benefit of other specific scopes as well.

On the contrary, if we make technology without a meta-project, it will be possible to accomplish useful things for a concrete goal but, being however limited, it will struggle to have an effective impact on socio-political transformation, so much so that it will not be frequently feasible because the system in which it is dynamized is instead synergistic with a contrasting meta-project. In this case, it may represent an opposing, degenerative opinion (trying to block the wave) that is susceptible to fail,[4] lacking a synergistic legal, economic and political framework. This course of action, which is only apparently comparable to having a meta-project instead, could suggest that the defended position is simply unrealistic and unfounded. A true self-goal. As soon as a battle starts in this way, even with heroic attitudes, one is lead to a defeat that remains and is added to future ones, so that the defended positions lose their credibility. We don't think it is useful to keep on feeding such a mechanism.

---

[4] I think this is the reason, or at least, one reason why degenerative projects are losers in the face of any progressive project. It is not just a psychological dimension of positivity, but the ability to integrate itself into the system efficiently.

From an ethical point of view, it is necessary to have a meta-project supporting the idea of the World we want, and for which we intend to develop a coherent technology. A meta-project that is *causa causarum* (final cause) (Stanford Encyclopedia of Philosophy 2006) of the technology and the various sciences, experimental and otherwise, dealing with AI. A meta-project that gives consistency to the teleological dimension of AI, still within the general teleological ethics as one of its applicative and instrumental scopes.

This will make possible a prudential governance of AI because it will clarify the end for which AI should develop as a righteous means, which seems to us to be the crucial point.

We propose below, as points for discussion, some notes that we consider useful for a possible meta-project for AI.

2.1 Note No. 1: It Must Be an Infinite Intentional Structure

The virtue of Prudence is considered the prime, constitutive cause of virtues in general. In the words of Josep Pieper, "Virtue is a perfect power of Man as a spiritual person; and justice, fortitude and temperance reach their perfection as the power of the whole Man only when they are based on prudence, that is on the perfect power of a just right conclusion; only through these perfect powers of determination instinctive inclinations toward the good are elevated into the sphere of spiritual decision of Man, from which truly human actions spring" (Pieper, 1959). This Virtue makes possible to recognize the fundamental goal of life in the concrete situation and -precisely because of this, I think—one is capable of choosing the appropriate means to achieve it. Still according to Pieper, "the resolution of Prudence is the extrinsic substantive form (measure) by which good action is what it is: only through the fundamental resolution of Prudence does it become right, strong, tempering" (Pieper, 1959). In our case, justice is composed in the regulatory norms, fortitude is expressed both in the robustness of systems (ability to resist) and resilience against external agents (ability to counterattack), and temperance can be declined, in an analogical sense, in the definition of even multiple Objective Functions, thus avoiding the form of "general purpose" (European Parliament, 2023). In this way, the prudential governance of AI allows us to responsibly choose goals and even evaluate them in their concrete applications, as well as in the dynamics of development according to its nature of machine learning. In this view, having awareness of the end we aim at with action and knowing why we act make our action a "perfect power", that is a truly good Human Act. Thus, the meta-project can give content and consistency to prudent action in all those particular sciences that contribute to its resolution, including AI.

Placing a possible meta-project in the prudential dynamic requires the end neither to be immediate nor isolated, and yet it must be simultaneously intelligible. In Spaemann's words, "to every question about 'why?' addressed to the acting Man, the answer is satisfactory only when it reveals an intentional structure accessible to us. We can understand the answer". At the same time though, and this is not a contradiction, "the answer is satisfactory only when the intentional structure is infinite (...) we can say: when its purpose is a purpose in itself, or when the normality that defines it is an absolute sense connection, stable in itself" (Spaemann & LöW, 2013).[5] Finally, if the *locus* of a meta-project is "prudential acting," then it will be appropriate to speak of a meta-project for those dealing with AI rather than risk to hypostatize AI as if it were an acting subject.

Hence, a first indication can be drawn to define a meta-project for AI practitioners: having an infinite intentional structure or with an absolute sense connection.

2.2 *Note No. 2: A "Science for Human Being"*

To speak of "Science for Human Being" is not a slogan, it instead represents a precise cultural program, a true anthropological stance that can be chosen as the root of coherent scientific and technological action. The starting anthropological position interprets science and technology as an intentional act of an acting human subject (Marcos, 2010). At the core of any scientific and technological action we will always find the HB with their awareness, intentionality, and capacity. In this way, we distance ourselves from a hypostatization of Science and any form of AI, which the HB is user, victim or spectator of. This vision, being applied to the concrete of the various practical/experimental sciences, makes easier not to reduce science to mere Methodology (which is why the Clinical Methodology we propose is rooted in Popper's Critical Rationalism in synergy with Evidence Based Medicine (EBM), the Personalism of the Lublin School, and the Ethics of Work Well Done (Tambone & Ghilardi, 2020)). It makes also possible to direct Research towards a median path between Reductionism and Anti-Reductionism (what we call Conscious and Cooperative Reductionism (Tambone & Ghilardi, 2012)), to try to overcome the view of the use of Big Data as mere correlation by recovering interpretation, enhance as much as possible the use of new technologies to be more effective and all within a systemic anthropological vision. In this way, scientific and technological action will never be anything neutral, "extra-ethical". In the words of St. Josemaria Escriva, "University knows that the necessary scientific objectivity rightly rejects all ideological neutrality, all ambiguity, all conformity, all cowardice: the love for truth engages the life and the entire work of the scientist, and keeps him firm in honesty in the face of uncomfortable situations, because this honesty of commitment does not always find a favorable echo in public opinion" (Escrivá de Balaguer, 1993).

Scientific action has its own inherent ethicality and, therefore, the first duty of a worker is to work well, whatever job they have. The awareness that the work well done is not the ultimate end but a partial end reinforces the need to accomplish it as perfectly as possible, since, in order for a partial end to be oriented towards the

[5] Translation provided by the authors.

ultimate end, it must first of all be a good partial end in itself. Its deconstruction would not allow its proper use, even in view of the end with an absolute sense. Let it be clear that this reference to the following end is not the dynamic of romanticism (Stanford Encyclopedia of Philosophy, 2017) that, by replacing the good with the sublime, makes it a desire never possible, but rather it is the dynamic of the completeness of meaning that, in human beings, tends to the first cause as the final cause (Hollencamp, 1948). This dynamic, as (vertical) relationship, not only gives meaning to the horizontal relationships that work entails toward individuals and towards the common good, but also substantiates the vocational dimension of personal work through the concept of participation (Fabro, 2005). The self-referentiality of the human act within this vision reveals how ethics applied to science and technology has personal happiness as its first goal, even before the effectiveness of work and the goodness of its consequences. In this way, we come to better understand the first dimension of Science for Human Being, which will be a professional action that perfects the HB who performs it in view of the common good of society.

But this is not enough to understand that this is a true anthropological stance.

Human nature in the Lutheran vision is, after original sin, corrupt: Luther states that human nature, after original sin, is totally corrupt, absolutely powerless to do any good, and identifies sin itself with concupiscence (Maritain, 1970). In this view, human beings in their actions will only corrupt and do evil, precisely because (corrupt) nature remains the principle of operations.

The consequences are very concrete. For example, within the evolutionist view the human species is the only one that does not adapt to the environment but adapts the environment to itself. This implies that it is the only species capable of causing an ecological problem. If it is a species of corrupt beings, the logical conclusion will substantiate the apocalyptic vision for which HB will increasingly ruin himself and nature, leading everything and everyone towards self-destruction. In this context, science would play a role in rational recovery, a sort of restoration of Aristotelian action that happens *kata logon*. However, it should develop autonomously, not so much for saving humanity (which, being corrupted, cannot be saved) but to go "Beyond Human Being," into a post-humanistic vision with Technological Singularity (Kurzweil, 2005) as the necessary evolutionist horizon. An "artificial *kata logon*" consistent with an Artificial *Logos* capable of transforming chaos into cosmos for a second time. Of course, this will be an artificial cosmos, a virtual reality. Science, therefore, is not for Man but for what goes Beyond Man, replacing him and becoming self-aware and auto-poietic: basically, it is a Science Beyond Man, an artificial general intelligence (AGI) Beyond Human Being.

In the Catholic view instead, human nature is wounded after original sin. After the creation and the fall, redemption constitutes a theological anthropology full of loving care towards the HB, who is called to sanctify himself as an image of God, His son in the Son (Second Vatican Council, 1965 [no. 22]). However, the Son is the eternal *Logos*, to the extent that the *imago Dei* in the HB emphasizes its rationality, always founding their freedom, which makes them capable of communion with creation, other human beings, and God Himself. In this perspective, science is a human act that can heal the various wounds of human nature, if done according to reason. Just as the Redeemer comes with His messianic action to heal every HB (divine physician), Man, united by Grace to God, performs theandric actions when acting wittingly by choosing the Good. Thus, scientific and technological action will become a resource for humanity, being of service to creation and every HB. It will be a Science for Human Being, an AI for Human Being.

Hence, a second indication can be drawn for governing AI: make the choice of an AI for Human Being, meaning an AI that remains a tool at the disposal of the HB and for the common good.

In addition to what has been said so far, the choice of a Science for Human Being is currently even more significant because it opposes the tendency to work for a Science for Humanity and not for the individual Human Being. This should concretize Asimov's "zeroth law," which, as we well know, inspires the Longtermism of Elon Musk and others (Chiusi, 2023).

2.3 Note No. 3: Restoring Normative Force to Natural Law through Ecology

In the "Republic", Plato structures the education path of the Guardians of the State (Plato et al., 2007) in relation to the general interest of the State and according to the characteristics he deems good for the State. In other words, he states something strategic yet simple: if we want to know how to educate the Guardians of the State, we must first know what kind of State we want. This lesson leads us to understand that if we want to change the world, we must know the world we desire. To put it in another way, with a proactive rhetoric, we want to have an objective with an infinite intentional structure as our horizon. In my view, the world we find desirable and that can provide a real orientation for personal and collective daily life is the Christian vision of the world. This choice does not mean taking a confessional reference necessarily experienced at the level of theological faith.[6] The essence of Christianity, being rationally shareable, can be accepted even by those with a different faith or who live a practical or theoretical atheism. This meta-cultural value of the Christian message implies that the Christian vision can be embraced by every form of culture and, for the same reason, cannot be reduced to a specific culture. In the words of Saint John Paul II, "the Gospel is not contrary to this or that culture as if, when they meet, The Gospel wanted to deprive that culture of what belongs to it and force it to assume extrinsic forms that are not conforming. On the contrary, the announcement that believers bring to the world and cultures is a real form of liberation from every disorder introduced by sin and a call to full truth" (Pope John Paul II, 1998 [no. 71]) at the same time.

[6] The stance taken by F. Capra is significant, as he has been advocating for several years that every scientific approach must have a transcendent reference that opens up to necessary cosmological interpretations.

On the other hand, those who share this goal will be aware that the way to achieve this transformation, even through a systemic vision of AI, can be a matter of debate and even diversity of opinions, often due to the necessary effort of *epikeia*, other times to the legitimate creativity forged in cultures of diversity, and still others due to communicative needs dictated by target diversity.

The foundation will still be the truth of the world that already exists as the goal of scientific knowledge (of various experimental and non-experimental sciences): we do not want to create a new world; we want the existing world to be more perfectly what it should be. And this is what the world tells us, in the first instance through a knowledge of its laws, which are, in the progression of laws, a reflection of cosmic law and a source of positive law. The world we want is the existing one, ecologically restored, a world in which ecology as well as the laws of nature have once again normative (Pope Francis, 2015; Pope Francis, 2023) force. The ecological vision we speak of should be nourished by a cosmological thought that can revive the philosophy of nature beyond an emerging pantheism. To this end, it will be important to recover the hermeneutics of knowledge by not confusing it, in the wake of Paul Ricoeur, with that of understanding (Luño, 2023). This ecological choice gains particular significance when thinking about a meta-project for AI, in relation to the Longtermism of (MacAskill, 2019) and (Bostrom, 2013), for whom the future of humanity must lead humans to become an interplanetary species and to colonize new planets, as ecological catastrophe seems inevitable. In harmony with the Science for Human Being, the ecological vision we refer to proposes a Science for the Earth, i.e., not a "human ecology without environment and without a planet" but an ecology that enhances the crucial role of humans as responsible custodians of the Earth and the environment.

Hence, a third indication can be drawn to define a Meta-project for those involved in AI: to restore normative force to natural law through an appropriate declination of integral Ecology (Pope Francis, 2015).[7]

2.4 Note No. 4: Making It Attractive—Pop-philosophy

As we know, Plato argues that those who give Speeches must first know the truths of the things they speak about, secondly they must know the (Plato et al., 2011). Understanding the Truth and the Method of knowledge seems to fall under the Note

[7] See No. 60: "Finally, we acknowledge that various views and lines of thought have developed regarding the situation and possible solutions. On one extreme, some argue at all costs for the myth of progress and assert that ecological problems will be solved simply with new technological applications, without ethical considerations or fundamental changes. On the other extreme, others believe that the human species, with any of its interventions, can only be a threat and compromise the global ecosystem, so it is advisable to reduce its presence on the planet and prevent any type of intervention. Possible future scenarios, as there is not a single solution. This would leave room for a variety of contributions that could engage in dialogue towards comprehensive answers".

No. 2 (Science for Human Being), while "knowing the souls of the people they address," that is understanding the deuteragonist, can give substance to this fourth note that I wish to propose. We can know the truth we want to support and govern society with, but if we cannot make it understandable and convincing, it doesn't serve much purpose.

To this end, it might be useful to consider that the transcendentals of being can be not only a solid anchor for logical coherence but also a good connection between truth and rhetoric, a suggestion of effectiveness. From a theoretical standpoint, we find this theme from Pythagoras to the present day. "What is intelligible is also beautiful," summarizes (Chandrasekhar, 1987), referring to Mathematics and other Sciences, considering beauty as a hermeneutic indicator of truth even before a hypothesis is tested. Beyond the speculative robustness of the convergence in the unity of transcendentals (Alvira et al., 1991, Alessi, 1992, Weisheipl, 1974, Thomas Aquinas, 2005), what seems useful to observe now is the modulating effectiveness that typical products of contemporary popular culture (social media, movies, music, comics, and videos) have as pleasant, desirable, and enjoyable carriers of content, even of technological interest. It doesn't matter if the concept is more or less difficult, what matters is making it understandable and enjoyable; in fact, the true and the good seem more true and good if they are also beautiful. This has been achieved many times and continues to be realized through what Deleuze defined as pop-philosophy in 1972 in his "Anti-Oedipus." It is about conveying theoretical content (in a broad sense) through products which are typical of "pop" culture and are effectively capable of reaching laypersons and spreading like music does to the masses. Deleuze thinks of Woodstock or the French May. The underlying idea feeds on the theory of intensities, where "there is nothing to explain, nothing to interpret, nothing to understand. It's a kind of electrical connection (Deleuze & Guattari, 2009). In other words, it is about using the estimative pathway (enhancing it) outside the critical/reflexive thinking. This connection is primarily emotional, hedonistic, and energetic in a more purely biochemical and erotic sense. These connections, precisely because of their rewarding characteristic, effectively produce dependencies. The reduction to the level of estimation involves a change of the personal evaluative key through the metamorphosis of consciousness into a pseudo-consciousness no longer anchored to a judgment of truth, but at the service of a judgment of pleasure. Faced with these phenomena, rational opposition is very ineffective as it appears, to the pseudo-consciousness, ridiculous and losing. Only a political dominion of passions capable of making us savor again how good (pleasant) the good and the true are can lead to a recovery of true personal freedom and, therefore, of the ability to love. The education of the very young and the therapy of various dependencies hinge on this point. We mean a political dominion of passions and, therefore, non a despotic mastery, that is to say aiming for a proactive pop-philosophy whose main goal is the recovery of critical/reflexive thinking and not just the contrast of the pop-philosophy of the estimative reduction. For an AI meta-project, this could mean developing a pop-philosophy that is not only dystopian, as it has been mainly done so far, creator of possible scenarios which are useful for market preparation, but rather innovative cultural interaction products that involve society in AI design. Such a bottom-up dynamic could be effective in creating greater trust in AI, as already investigated for social (Campanozzi et al., 2019) robotics, perhaps using the UX (User Experience) methodology (Nielsen Norman Group, 1998).

The critical/reflexive approach resulting from this cultural strategy should be able to engage with the two giants of pop-philosophy related to technology and AI: Philip Dick and Isaac Asimov. It is a challenge to take seriously the questions they pose as fundamental for governing generative technology.

Philip Dick's question is embodied by the Android Hunter (Dick, 2010), and it is anthropological, technological, and epistemological at once. The Android Hunter must destroy androids without killing humans, but here is the question: "how to distinguish a perfect android from a real HB, and what makes a HB human"? The question becomes even more dramatic when it comes to be reflective: "how can the Android Hunter be sure he is not an android himself"? The AI Meta-project has an important and fascinating critical/reflexive goal in this question: to develop a critical/reflexive Humanism that seeks what is specific to the HB beyond their functions and appearances, an unexpected revaluation of substance that should lead to an equally unexpected rediscovery of the concept of the "soul" through interaction with generative AI.

If Philip Dick's question is mainly anthropological, Isaac Asimov's is twofold, as anthropological and ethical. The anthropological question is embodied in Psychohistory, representing the *fil-rouge* of the Foundation Cycle (Asimov, 1991). It can be considered as the pop-philosophical narrative of Hilbert's Assiomatization project, the desire to govern the world through a Mathematics capable of predicting the behavior of Humanity as a collective organism. This anthropological/political determinism, that inspires contemporary projects like those of Elon Musk, could be reinterpreted in light of the story of Hilbert's failure thanks to Gödel's incompleteness theorem and Heisenberg's indeterminacy (Israel, 2009). On the other hand, the ethical question is effectively proposed through the four laws of Robotics (Asimov, 1991), which still today serve as an operational reference for debates concerning the ethics of new technologies. In this case, critical reflection can lead us beyond Asimov's laws, especially the zeroth law, thus recovering the value of the single HB compared to an overly "massifying" and concretely nonexistent concept of "Humanity."

Other possible paths to consider concern the new welcoming Cozypunk, Hopepunk, and Solarpunk science fiction, where the utopian rather than dystopian dimension seems to prevail. In this context, AI has a positive role within a positive society cultivating positive values such as love, kindness, and inter-species collaboration (Chambers, 2022), in a world where what seems to matter is an apparent normality through which a new naturalism emerges that is only apparently minimal.

In summary, being aware that AI is a tool and not an end makes us desire its prudent governance within a meta-project based on the four notes that we propose. This could revitalize our vision of good and evil and a humanistic insight that rediscovers pre-notions such as "nature" and "soul".

3 Prudential Government & Universal Government: The United Nations' Proposal

What has been discussed may seem in line with the United Nations' proposal for a Universal Governance of AI found in the Interim Report: Governing AI for Humanity (United Nations, 2023). In order to understand similarities and differences, it will be useful to reflect on the first three guiding principles proposed by the Advisory Body.

3.1 Guiding Principle 1: AI Should Be Governed Inclusively, by and for the Benefit of All (United Nations, 2023 [No. 46–47])

The first proposed principle aims for inclusivity in the governance process of AI, so that "All countries, big or small, must be able to participate in AI governance" (United Nations, 2023 [46]). The applications of a powerful tool like AI can greatly vary and should be implemented based on the real needs of different socio-political scenarios. The needs in Los Angeles will be very different from those in Rome or Beijing. The educational challenges that AI can help address in different parts of the world are a far cry from the security applications it may need to perform in a country at war. Developing generative AI programs might be the main purpose for a digitized population, while remote diagnosis systems and tele-medicine services would be the right choice for an area with poor access to healthcare services. Perhaps for this reason, the first guiding principle seems to open up to the responsibility of local governance, regardless of its socioeconomic power. Therefore, it would be a Universal Governance which is inclusive of Local Governments, that is, we feel compelled to say, of a Universal Governance of the technological dimension in the service of the real political choices of Local Governments. This could be an apparently acceptable approach if the training of AI were not actually a political choice.

As we know, AI development has a learning phase (Russell & Norvig, 2016) that presupposes not so much an "a priori" knowledge but rather a knowledge starting from the transfer of data to be realized through processes of induction and deduction aimed at supervised, unsupervised, and reinforcement learning.[8] Each phase of this process can be carried out in very different ways resulting from decisions made by trainers and based on the goals they set. In this way, even if the product appears "neutral" with a technological identity being "impartial", it will be the result of a planning set by a real interest group. For this reason, the progress we are observing within AI is not "neutral" but obviously guided by those who invest, develop, and

[8] Training is a complex process with stages outlined as follows: (1) problem definition; (2) data collection and processing; (3) formulation of training sets; (4) model selection (with hyper-parameter optimization); (5) model training; (6) validation set to refine hyper-parameters; (7) evaluation on a test set; (8) actuation and monitoring.

commercialize. The political nature of the training phase becomes even clearer in light of the concept of the "incommensurability of the Human," which for McIntyre is also manifested through dialectical argumentation, conceptual frameworks, linguistic practice, symbolism, historical contextualization, and social reference functions. Language is linked to and arises from rationality, it connects with beliefs, changes with action and through relationships (McIntyre, 2009).

We think inclusivity should start, in this case, from respecting diversities and building anthropologically sustainable relationships that go beyond the principle of overpowering, in order to build the common good. In this case, inclusivity doesn't necessarily have to be at design or usage level (which is impossible in the current state of affairs) but at the level of building well-being through AI. People have different cultural levels and abilities, and this diversity should be respected: no one can propose as a goal that everyone acquires computer skills adequate to program intelligent systems; on the other hand, no one can assert that someone with culinary skills is useless in a highly technological world. Artists must have the opportunity and freedom to cultivate their genius and do things beyond an algorithm, such as a poem or a painting. Anthropologically sustainable relationships are those that implement the necessary conditions for everyone to achieve their ultimate goal, which is the content of the true common good. This respect for diversities and relational goods stems from conceiving the HB as an end (Chalmeta, 2003), once again emphasizing their ontological difference from technologies in general, and AI in particular, which will rather be tools for Human Being.

3.2 *Guiding Principle 2: AI Must Be Governed in the Public Interest (United Nations, 2023)*

What has just been said about the necessary balance between Universal Governance and Local Governance is valid for this second guiding principle. However, it suggests at least two other considerations. Firstly, the approach towards the industrial world appears implicitly distrustful (United Nations, 2023),[9] implying that private interests are in conflict with the public interest (societal interest). This may be true in some cases but, at the same time, there are situations where it is quite the opposite due to the political-social structure of some nations. Secondly, this guiding principle

[9] See N. 48. The development of AI systems is largely concentrated in the hands of technology companies. (…) N 49. As shown by the experience with social media, AI products and services can scale rapidly across borders and categories of users. For this reason, as well as wider considerations of opportunities and risks, AI must be governed in the broader public interest. "Do no harm" is necessary, but not sufficient. A broader framing is needed for accountability of companies and other organisations that build, deploy and control AI as well as those that use AI across multiple sectors of the economy and society across the lifecycle of AI. This cannot rely on self-regulation alone: binding norms enforced by member states consistently are needed to ensure that public interests, rather than private interests, prevail.

should manifest the will to ensure freedom of research even in the technological field. This is true if we want to refer to the model that outlines the phases of research in "independent scientific research/finalized research/development," at least concerning the first two phases of researchers' work. The second guiding principle could effectively be applied to the third phase, the development, meaning the application distribution of the results of the first two phases, thus ensuring the freedom of scientific research and the control over the social impact of its possible applications.

3.3 *Guiding Principle 3: AI Governance Should Be Built in Step with Data Governance and the Promotion of Data Commons (AI Advisory Body, 2023 [No. 51])*

The third guiding principle explicitly highlights the real connection between the enormous power of AI and the significant power associated with the ownership of Big Data. Specifically, the great importance of Big Data passes through what we can call personalization, Big Data analytics, and intelligence. As we know, the use of personalizing algorithms on a commercial level, for example, serves to promote product purchases through social messages tailored to the personal feedback of digital tool users. In the words of Bezos, "We see our customers as invited guests to a party, and we are the hosts. It is our job every day to make every important aspect of the customer experience a little bit better" (Suite Feedback, 2019). Analytical use of Big Data aims to better understand its engagement target, improve organization and operational processes to increase efficiency, and make the right strategic decisions at the right time. This is no longer about modulating the behavior of a potential end-user, but understanding the predictable dynamics regarding the efficiency of one's actions. Finally, on the subject of Intelligence, it involves collecting information considered useful for State security. As can be seen, this "Triad" fits into a theater of generic conflictuality that can be characterized, time after time, as a dynamic of economic/financial market conflict, military conflict, or national security. In all these cases, global sharing of Big Data appears utopian and difficult to comprehend, while it would become a significant goal in a framework of universal peace and cooperation.

In summary, for the UN proposal to be in line with the Prudential Governance we are discussing, the first three guiding principles should ensure that:

1. Universal Governance is limited to the technological dimension and serves the real political choices of Local Governments.
2. The freedom of scientific research is respected also in the technological field, ensuring that its applications are always in the service of the common good.
3. The sharing of Big Data becomes a realistic goal within political dynamics of security and peace.

4 Prudential Governance of AI and the Common Good

We define the common good as the set of conditions of social life allowing both groups and individual members to achieve their perfection more fully and expeditiously (Second Vatican Council, 1965 [no. 75]). Connecting AI governance to this vision of the common good allows us to refer to the 17 Sustainable Development Goals proposed by the United Nations (2023), thus going beyond the overgeneralization of Asimov's Zeroth Law for Robotics. The 17 SDGs can indeed be related to four goals particularly important for New Technologies that want to be prudentially governed, namely: Bio-safety, Bio-cooperation, Bio-synergy, and Hybrid Bio-symbiosis.

4.1 *Bio-safety*[10]

Bio-safety[11] refers to the virtue of Prudence and the Precautionary Principle. The commitment to move from classic Risk Governance to what we call "Systemic Bio-Risk Management" embraces in the usual doctrine the vision of Complexity Theory. This theory asserts that the multi-causality of the effect and the feedback produced, trigger a circular causality that is not limited to the system described herein because, since there are no closed systems, it is also influenced by inter-systemic connections that chain together progressively and never conclusively. In this risk management system, both the correlation and interpretation become the tool fed by the maximum number of possible data, which are then interpreted by dedicated algorithms. Being aware of what has just been said implies that, precisely because of this inter-systemic correlation, the data concerning risk management are not only those traditionally understood as sensitive. In this logic, all data will be sensitive, and their usefulness for risk management will depend only on the computing power of the available algorithm and the ability to collect data. Bio-safety becomes, in this perspective, the center of a concept of global security, since if we are interconnected without losing our individuality the security of the acting subject regarding the risk management is

[10] The concept of Biosafety is connected to the following Sustainable Development Goals (SDGs):

- G2: End hunger, achieve food security and improved nutrition and promote sustainable agriculture.
- G6: Ensure availability and sustainable management of water and sanitation for all.
- G10: Reduce inequality within and among countries.
- G11: Make cities and human settlements inclusive, safe, resilient and sustainable.
- G16: Promote peaceful and inclusive societies for sustainable development, provide access to justice for all, and build effective, accountable and inclusive institutions at all levels.

[11] Bio-safety refers to the measures and practices adopted to manage risks of any kind related to "bios", life, within the One Health perspective but also with an increasingly articulated projection into digital and virtual worlds.

central for the security of animal, plant, and even prebiotic systems. This is a splendid reflection of a highly effective One Health Human-Centered logic.

To this end, we consider it important to enrich Systemic Risk Management with the dimension of New Technologies, and with a dedicated Cloud too.

Bio-cooperation[12] Bio-cooperation,[13] within a multidimensional and systemic vision, will not only refer to recover humans from the hardest or most dangerous jobs, but making work itself easier and safer. Moreover, this technological Bio-cooperation will also be developed as plant and animal cooperation, so that the working relationship that requires the interconnection of these different vital dimensions is increasingly integrated, so as to create, in synergy with the systemic concept of One Health, a "Systemic One Job" that would present itself as a remarkable development of the supply chain logic. This type of extra-human technological cooperation will allow a prudential human governance of the animal and plant world through AI, naturally with Delegation Standards and the recovery of valuable feedback to be connected to the goals just described for Bio-safety.

For this purpose, an experimental body will be an essential tool to serve as a place for critical thought where drawing up shared "Good Technological Practice" similar to the Good Clinical Practice that has regulated clinical trials for decades. Such work will require cooperation with the industrial sector.

Bio-synergy[14] The term "bio-synergy" is not widely used or present in academic or scientific literature. There is neither consolidated nor recognized definition for the concept of "bio-synergy." This term might have been created or used in specific contexts or particular sectors without a universally accepted definition. An agreed-upon definition may need to be proposed in the future.

---

[12] Related SDGs:+

- G4: Ensure inclusive and equitable quality education and promote lifelong learning opportunities for all.
- G5: Achieve gender equality and empower all women and girls.
- G7: Ensure access to affordable, reliable, sustainable and modern energy for all.
- G8: Promote sustained, inclusive and sustainable economic growth, full and productive employment and decent work for all.
- G9: Build resilient infrastructure, promote inclusive and sustainable industrialization and foster innovation.

[13] The concept of Bio-cooperation requires, for the definition of cooperation as "work done for or with others for the realization of an enterprise or the achievement of a goal," the definitive recovery of the principle of individuation and the identification of the goal, which, in line with what has been said for Bio-safety, will be the Common Good. It is about developing AI increasingly better so that its use by humans for work and social life ensures safety, effectiveness, efficiency, etc.

[14] Related SDGs:

- G1: End poverty in all its forms everywhere
- G3: Ensure healthy lives and promote well-being for all at all ages.

However, by analyzing the meaning of the components of the term, we can provide a first general interpretation. "Bio-" refers to "life" or "biology," while "-synergy" indicates cooperation or interaction where the overall result is greater than the sum of the individual parts. Therefore, it might be inferred that "bio-synergy" could refer to a concept in which biological components or life elements interact synergistically, achieving results or effects superior to those that would be obtained by treating such components separately.

The "Good Technical Practice" outlined above will play an important role in regulating the experimentation aiming at Human-AI Bio-synergy, intending it not as the activation of the psycho-emotional perception of the "lived moment," typical of a psycho-dynamic interpretation of the bio-synergy concept, but as "combined and simultaneous action, collaboration, cooperation of multiple elements in the same activity, or for the achievement of the same purpose or result, which involves a greater performance than that obtained from various separate elements."[15] In our case, the synergy of AI with a human is in the first instance particularly important to fill the functional gap produced by congenital disabilities or those resulting from various types of trauma. To this end, it will be useful to define the principle of operations to reinforce the principle of individuation. The needed tool will be an experimental body to be developed in cooperation with entities that have significant clinical records regarding disabilities, rehabilitation, and workplace injuries.

Hybrid Bio-symbiosis[16] "Bio-symbiosis" is a term that combines "bio" (referring to life or biology) and "symbiosis," which alludes to a close and lasting relationship between two organisms of different species, both beneficiaries from mutual collaboration.

Bio-symbiosis could be interpreted as a form of interaction or cooperation between different biological organisms or systems, where a symbiotic bond occurs leading to mutual benefits, well in line with the idea of the common good we refer to. In this context, the interaction between organisms implies an exchange of resources, energy, and information contributing to the well-being and survival of both organisms involved.

Applying the concept of Bio-symbiosis to the interactive link between a living being and an AI reality is crucial as an interpretative and evaluative framework for

[15] See https://www.treccani.it/vocabolario/sinergia/ (accessed May 12, 2024).

[16] Related SDGs:

- G12: Ensure sustainable consumption and production patterns.
- G13: Take urgent action to combat climate change and its impacts.
- G14: Conserve and sustainably use the oceans, seas and marine resources for sustainable development.
- G15: Protect, restore and promote sustainable use of terrestrial ecosystems, sustainably manage forests, combat desertification, and halt and reverse land degradation and halt biodiversity loss.
- G17: Strengthen the means of implementation and revitalize the Global Partnership for Sustainable Development.

Neuralink experimentation.[17] In this case, in addition to patient safety and knowledge of transient or permanent effects on the human subject (considering, for example, reverse plasticity), an anthropological/ethical consideration will be valuable on both the principle of individuation and the concept of nature as a principle of operations. Therefore, this concept has a strong ecological significance but for a hybrid ecology, and precisely for this reason, it is of great importance. We mean that the Bio-symbiosis we are interested in is a Hybrid Bio-symbiosis that makes us more aware of how the ecological vision may need to expand to technological products entering into symbiosis with living beings, thereby shaping a hybrid ecology.

5 Continuation...

There are still numerous ethical themes suggested by AI that need exploration. Still, for now, it seemed useful to focus on the possible usefulness of developing a prudential governance of AI design and usage.

The prudential dynamic of means and ends has to deal with the enormous development that, since the last century, has produced an increasingly capable technology of previously unthinkable objectives. Basically, this has led to the subversion of the strategic calibration of means relative to ends (Kissinger et al., 2023). New technologies, particularly AI, are producing a new strategy of international competition, challenging that generated by nuclear weapons. For some analysts, AI introduces a post-Enlightenment era that by changing, at least seemingly, the possibility of consequentialist calculation, necessarily redefines the rules of any "social contract," casting doubt on the ownership of necessary information.[18] If the introduction of non-human logic into decision-making processes, that is when choosing the goal, becomes not analogical but symbiotic or even substitutive, then there is the most radically anti-prudential subversion, that is the replacement of the end by the means. Therefore, the instrumental dimension of technologies, AI in particular, requires the recovery of a humanistic awareness regarding the essential characteristics of the HB and a clear choice concerning a "Science for Humanity." Such a choice, far from being a mere slogan, has a decisive substance that can be, in our case, that intentional infinite form we need.

[17] See https://neuralink.com/ (accessed May 12, 2024).

[18] Kissinger, H., Schimdt, E. & Huttenlocher, D. (2021). The Age of AI: And Our Human Future. Little, Brown and Company.

References

AI Advisory Body. (2023, December). *Interim report: Governing AI for Humanity*, n. 13. Retrieved May 11, 2024, from https://www.un.org/sites/un2.un.org/files/ai_advisory_body_interim_report.pdf

AI Safety Summit. (2023, November 1–2). *The Bletchley declaration by countries attending the AI safety summit*. Retrieved May 13, 2024, from https://www.gov.uk/government/publications/ai-safety-summit-2023-the-bletchley-declaration/the-bletchley-declaration-by-countries-attending-the-ai-safety-summit-1-2-november-2023

Alessi, A. (1992). *Metafisica*. LAS.

Alvira, T., Clavell, L., & Melendo, T. (1991). *Metaphysics*. Sinag Tala Publishers.

Aquinas, T. (2005). *Disputed questions on the virtues*. Cambridge University Press.

Arendt, H. (2005). *The promise of politics*. Schocken.

Aristotle. (2009). In L. Brown (Ed.), *The Nicomachean ethics*. OUP.

Asimov, I. (1991). *The complete Isaac Asimov's: Foundation series*. Spectra Books.

Baccour, E., et al. (2021). Pervasive AI for IoT applications: A survey on resource-efficient distributed artificial intelligence. *IEEE Communications Surveys & Tutorials, 24*(4), 2366–2418.

Baradaran, A. (2024). Towards a decolonial I in AI: Mapping the pervasive effects of artificial intelligence on the art ecosystem. *AI & Society, 39*, 7–19.

Bodeus, R. (1993). *The political dimension of Aristotle's ethics*. State University of New York Press.

Bostrom, N. (2013). Existential risks prevention as global priority. *Global Policy, 4*(1), 15–31.

Campanozzi, L. L., Guglielmelli, E., Cella, E., Ghilardi, G., Michilli, M., & Tambone, V. (2019). Building trust in social robotics: A pilot survey. *IEEE Technology and Society, 38*(4), 45–54.

Capital Economics. (2024). *AI, economies and markets–How artificial intelligence will transform the global economy*. Retrieved May 11, 2024, from https://www.capitaleconomics.com/ai-economies-and-markets-how-artificial-intelligence-will-transform-global-economy?_gl=1*189806n*_up*MQ..&gclid=EAIaIQobChMI_Jm7yIS8hAMVB6WDBx2b8wStEAAYAyAAEgKyPPD_BwE#economic-revolution

Cazzaniga, M., et al. (2024). *Gen-AI: Artificial intelligence and the future of work*. IMF Staff Discussion Note SDN2024/001. International Monetary Fund.

Chalmeta, G. (2003). *Introduzione al personalismo*. EDUSC.

Chambers, B. (2022). *A prayer for the crown-shy*. Tor Books.

Chandrasekhar, S. (1987). *Truth and beauty: Aesthetics and motivations in science*. University of Chicago Press.

Chiusi, F. (2023). *L'Uomo che vuole risolvere il Futuro*. Bollati Boringheri.

Deleuze, G., & Guattari, F. (2009). *Anti-Oedipus: Capitalism and schizophrenia*. Penguin Publishing Group.

Dick, P. (2010). *Do androids dream of electric sheep?* Gateway. Originally published 1968.

Ebbert-Streib, F. (2021). From the digital data revolution toward a digital society: Pervasiveness of artificial intelligence. *Machine Learning and Knowledge Extraction, 3*(1), 284–298.

Escrivá de Balaguer, J. (1993). El compromiso de la verdad. In J. Escrivá de Balaguer (Ed.), *Discursos sobre la Universidad* (pp. 106–107). Eunsa.

European Commission. (2021a, April). *Communication on Fostering a European approach to Artificial Intelligence*. Retrieved May 11, 2024, from https://digital-strategy.ec.europa.eu/en/library/communication-fostering-european-approach-artificial-intelligence

European Commission. (2021b). *Coordinated Plan on Artificial Intelligence*. Retrieved May 11, 2024 from https://digital-strategy.ec.europa.eu/en/policies/plan-ai

European Commission. (2024a). *AI Act*. Retrieved May 11, 2024 from https://digital-strategy.ec.europa.eu/en/policies/regulatory-framework-ai

European Commission. (2024b). *Artificial Intelligence in the European Commission (AI@EC) Communication*. Retrieved May 11, 2024, from https://commission.europa.eu/publications/artificial-intelligence-european-commission-aiec-communication_en

European Parliament. (2021, September). *China's ambitions in artificial intelligence*. Retrieved May 11, 2024, from https://www.europarl.europa.eu/RegData/etudes/ATAG/2021/696206/EPRS_ATA(2021)696206_EN.pdf

European Parliament. (2023, March). *General-purpose artificial intelligence*. Retrieved May 11, 2024, from https://plato.stanford.edu/entries/aristotle-causality/#ExplPrioFinaCaus

European Union. (2024). Regulation (EU) 2024/1689 of the European Parliament and of the council of 13 June 2024 laying down harmonised rules on artificial intelligence and amending regulations (EC) no 300/2008, (EU) no 167/2013, (EU) no 168/2013, (EU) 2018/858, (EU) 2018/1139 and (EU) 2019/2144 and directives 2014/90/EU, (EU) 2016/797 and (EU) 2020/182 (artificial intelligence act). *Official Journal of the European Union, L series, 1–144*. 12.7.2024.

Fabro, C. (2005). *La Nozione di Partecipazione*. EDIVI.

Filippou, M. C. et al. (2022). *Pervasive artificial intelligence in next generation wireless: The Hexa-X project perspective*. Retrieved May 11, 2024, from https://eprints.sztaki.hu/10403/1/Filippou_1_33191533_ny.pdf

Georgieva, K. (2024, January 14). *AI will transform the global economy. Let's make sure it benefits humanity*. International Fund Monetary Blog. Retrieved May 11, 2024, from https://www.imf.org/en/Blogs/Articles/2024/01/14/ai-will-transform-the-global-economy-lets-make-sure-it-benefits-humanity

Government of India–Ministry of Electronics & Information Technology. (2023, October). *India AI 2023*. Retrieved May 11, 2024 from https://www.meity.gov.in/writereaddata/files/IndiaAI-Expert-Group-Report-First-Edition.pdf

Hokusai, K. (1830–32). *Under the wave of Kanagawa*. Retrieved May 11, 2024, from https://www.metmuseum.org/art/collection/search/45434

Hollencamp, C. (1948). Causa Causarum (I). *Laval théologique et philosophique, 4*(1), 77–109. https://doi.org/10.7202/1019802ar

Israel, G. (2009). *The world as a mathematical game*. Birkhäuser Verlag AG.

Kissinger, H., Schidt, E., & Huttenlocher, D. (2023). *L'Era dell'Intelligenza artificiale*. Mondadori.

Kumar, A., Braud, T., Tarkoma, S., & Hui, P. (2020). Trustworthy AI in the age of pervasive computing and big data. In *2020 IEEE international conference on pervasive computing and communications workshops (PerCom workshops)* (pp. 1–6). Austin.

Kurzweil, R. (2005). *La singolarità è vicina*. Apogeo.

Kurzweil, R. (2006). *The singularity is near*. Penguin Publishing Group.

London, J. (1911). *A royal sport*. : 1980 editions.

Luno, A. R. (2023). Superamento del soggetto come coscienza ed etica della virtù. In V. Paglia (Ed.), *Etica Teologica della Vita* (pp. 157–163). LEV.

MacAskill, W. (2019). *Longtermism*. Effective Altruism Forum.

Marcos, A. (2010). *Filosofia dell'agire scientifico*. Academia Universa Press.

Maritain, J. (1970). *Three reformers*. Praeger.

McIntyre, A. (2009). *Dependent rational animals*. Bloomsbury Academic.

Nemitz, P. (2018). Constitutional democracy and technology in the age of artificial intelligence. *Philosophical Transactions of the Royal Society, A.3762018008920180089.*

Nielsen Norman Group. (1998, August). *The definition of user experience*. Retrieved May 12, 2024, from https://www.nngroup.com/articles/definition-user-experience/

Pieper, J. (1959). *Prudence*. Pantheon Books Inc.

Plato. (2011). In H. Yunis (Ed.), *Phaedrus*. Cambridge University Press.

Plato. (2007). In M. Lane & D. Lee (Eds.), *The republic*. Penguin Classic.

Pope Francis. (2015). Encyclical letter *Laudato si'* of the holy father Francis on care for our common home. Retrieved September 5, 2024, from https://www.vatican.va/content/francesco/en/apost_exhortations/documents/20231004-laudate-deum.html

Pope Francis. (2023). Apostolic exhortation *Laudate Deum* of the holy father Francis to all people of good will on the climate crisis. Retrieved September 5, 2024, from https://www.vatican.va/content/francesco/en/apost_exhortations/documents/20231004-laudate-deum.html

Pope John Paul II. (1998). Encyclical letter *Fides et Ratio* of the supreme Pontiff John Paul II To The Bishops of the Catholic Church on the relationship between faith and reason. Retrieved September 5, 2024, from https://www.vatican.va/content/john-paul-ii/en/encyclicals/documents/hf_jp-ii_enc_14091998_fides-et-ratio.html

Russell, S., & Norvig, P. (2016). *Artificial Intelligence: A modern approach* (4th ed.). Pearson Education.

Second Vatican Council. (1965). Pastoral constitution on The Church In The modern world *Gaudium et Spes* promulgated by his holiness, Pope Paul VI On December 7, 1965. Retrieved September 5, 2024, from https://www.vatican.va/archive/hist_councils/ii_vatican_council/documents/vat-ii_const_19651207_gaudium-et-spes_en.html

Spaemann, R., & LöW, R. (2013). *Fini Naturali* (C. Ruini, Trans.). Ares.

Standford Encyclopedia of Philosophy. (2017, June). *August Wilhelm von Schle*gel. Retrieved May 11, 2024 from https://plato.stanford.edu/entries/schlegel-aw/

Suite Feedback. (2019, June 12). *Amazon & customer experience: 13 Quotes from Jeff Bezos*. Retrieved May 12, 2024, from https://suitefeedback.com/amazon-customer-experience-13-quotes-from-jeff-bezos/

Tambone, V., & Ghilardi, G. (2012, May). Aware and cooperative reduction. *La Clinica terapeutica, 163*(3), e133–e143.

Tambone, V., & Ghilardi, G. (2020). *Philosophy and deontology of medical practice*. SEU.

United Nations. (2023, December). *Interim report: Governing AI for humanity*. Retrieved May 12, 2024, from https://www.un.org/en/ai-advisory-body

United Nations. Department of Economic and Social Affairs. (2023). *The 17 goals*. Retrieved May 12, 2024 from https://sdgs.un.org/goals

Weisheipl, J. A. (1974). *Friar Thomas D'Aquino: His life, thought and work*. Doubleday.

Beyond Digital Escapism

Roger Strand

Abstract At least two discourses remain powerful in contemporary Northern and Western societies and live in parallel in our academic institutions. The "Geos" discourse of the Anthropocene, supported by the system sciences, warns against the unprecedented magnitude of human activity and the resulting ecological damage. The "Bios" discourse of the Digital Revolution, fuelled by technoscience and industry, imagines wonderful new and virtual worlds. Some—the European Commission, for example—believe in the happy marriage of these two discourses, imagining a twin digital and green transition to sustainability. In these times, what duties should we attend to as public intellectuals? I argue for the importance of hope. Respecting the scientific fact of limits to growth, however, false hopes should be rejected. To the extent that hope has to be connected to an idea of progress, it would be seen to lie in other domains than those of material consumption, wealth or power. Hope should be sought in the moral and spiritual domains, counter-acting polarization, greed and violence by finding, showing and walking avenues to connectedness, love and compassion with the humans and other living beings of our beautiful planet. Recalling that information and communication technologies are technologies, the question would then not so much be whether they can be described as part of the human environment, but if and how they can be employed along such avenues, in service to the beings of the real world.

Keywords Anthropocene · Digitalization · Imaginaries · Imagination

---

R. Strand (✉)
University of Bergen, Bergen, Norway
e-mail: roger.strand@uib.no

A. Fabris, S. Belardinelli (eds.), *Digital Environments and Human Relations*,
Human Perspectives in Health Sciences and Technology 150,
https://doi.org/10.1007/978-3-031-76961-0_9

1 Introduction: A Crisis of Imagination[1]

At least two discourses remain powerful in early twenty-first century Northern and Western societies and live in parallel in our academic institutions. The discourse of the Anthropocene, fuelled by a growing body of scientific knowledge from ecology, climate research and other system sciences, describes the unprecedented magnitude of human activity and the resulting damage to the animal and plant species, natural habitats, and the integrity of nutrient cycles and ecosystems on our planet. It warns against the potentially devastating consequences for the human species if and when the crucial tipping points of the natural environment have been passed. The discourse of the Digital Revolution, on the other hand, fuelled by technoscience and enterprises within the triple helix of academe, industry and government, imagines new, virtual worlds that unfold opportunities for entertainment, job creation, wealth, entertainment, well-being and indeed new forms of human life and existence. Some—the European Commission, for example—even believe in the happy marriage of these two discourses, whereby the technological unfolding of a digital world will leave the physical world clean and green again, finally uncoupling growth from resource use and environmental damage.

In this context, there are several tasks that call for the services of philosophers. How might human lifeworlds and the very human condition change as twenty-first century individuals spend ever more time "in digital environments", that is, interacting with computer devices and the network of information and communication technologies (ICTs)? Taking ecology in its broadest sense, as the study of what there is in the house (*οἶκος*), does human ecology change when smart phones have replaced TV sets and dogs as family members? All of these questions are important and raise their own ontological, epistemological and ethical issues.

This book chapter, however, takes a different route. Rather than diving into the implications of novel interactions between human individuals and ICTs as material objects, I shall reflect on a different (albeit related) phenomenon, namely the collective *imagination* being exerted around the digital. In particular, I am interested in the relationship between the two science-fuelled discourses mentioned above—the one of the Anthropocene and the environmental crisis, and the one of the prospects of a bright digital technological future. Let me also state my limited point of departure: Being particularly worried about environmental degradation, I am going to ask whether the discourse of the Digital Revolution carries the risk of political, scientific and civic escapism and negligence of environmental degradation and the loss of what a biologist would call the *real* ecology.

As such, even if I consider this chapter to be a philosophical contribution, I should warn the reader that there will be little philosophical sophistication in what

---

[1] I am indebted to Adriaan Eckels for the term «crisis of imagination», and even more so for being introduced his important work at the European Commission-Joint Research Centre in Ispra, Italy, building bridges between science and art as a response to that crisis, see https://science-art-society.ec.europa.eu/naturarchy-0#naturarchy-towards-a-natural-contract (accessed May 12, 2024).

follows. This is an easy read. I shall not discuss sophisticated theory, present refined conceptual analysis or expose subtle logical flaws. Rather, the notion of philosophy being practised in this chapter, is closer to Deleuze's idea of stuttering (Deleuze, 1994) or even simply what in everyday language would be called philosophizing, reflecting, re-interpreting and dwelling upon a possible shared understanding of our world. Accordingly, it may be that this text will not age well. Perhaps its philosophical relevance fades within a few years, and its potential interest only lies in being historical evidence of past troubles and concerns and apparent attempts of positing self-defeating prophecies. In fact, I hope so, because the troubles and concerns that will be discussed, do seem severe.

There are rich academic traditions on the understanding of the social, political and scientific importance of imagination and imaginaries, going back to Gaston Bachelard and Hannah Arendt and through Cornelius Castoriadis, Benedict Anderson and not the least Charles Taylor. I will not review this body of scholarship but simply state as my basic assumption that collective imagination indeed plays an important role in shaping social institutions and practices. In particular, Sheila Jasanoff and her co-workers have shown how sociotechnical imaginaries, understood as collective visions of desirable future social and technical order, are instrumental in the coproduction of science, technology and society (Jasanoff & Kim, 2015). Imagination can as such be seen as a means for or a type of "future-making", or in less metaphorical terms, as a type of practice that helps define goals, plans and actions both on the individual and collective level. This is particularly true in the field of research and innovation, which by its nature aims at producing something (immaterial or material) that did not exist before and perhaps was also inconceivable before.

Imagination is in that sense an important if not indispensable cultural resource. Critiques of imagination—at least those informed by the above-mentioned scholarship—do not object to imagination being exerted or deployed as such but rather to the particular features of the case of imagination at hand. Jasanoff has shown how sociotechnical imagination may be narrowly limited or biased by relying too much on the desires and worldviews of a narrow circle of actors (typically members of a scientific and technological elite that invest identity and capital into a certain scientific or technological domain). For example, one critique of contemporary sociotechnical imagination is that it relies too heavily on the worldviews, goals and desires originating in biotechnology and life science, at the expense of other understandings of humanity and its place in nature (Jasanoff, 2019).

In an attempt to understand the visions of precision medicine, I distinguished between imagination and imaginaries on one hand, and *fantasies* on the other (Strand, 2022). My use of these terms was simply that of common sense. For instance, the Merriam Webster online dictionary defines "imagination" as "the act or power of forming a mental image of something not present to the senses or never before wholly perceived in reality",[2] while "fantasy" is defined as "the power or

[2] See https://www.merriam-webster.com/dictionary/imagination (accessed May 12, 2024).

process of creating especially unrealistic or improbable mental images in response to psychological need" or the image created by that process.[3] Put simply, fantasy is a subset of imagination, qualified by creating unrealistic or improbable images.

Now, the whole point with sociotechnical imagination is to create images of states of affairs (social and technological orders) that do not exist yet—say, a society fueled by inexpensive energy produced by nuclear fusion reactors. The mere fact that fusion reactors do not exist yet, that we do not have sufficient scientific knowledge to build or even design them, and that a lot of research has been made with arguably slow progress, does not make fusion reactors a fantasy. This can be contrasted with the vision of making a perpetual motion machine, that is, a machine that can work forever without external energy supply. The perpetual motion machine is a fantasy. The difference between the fusion reactor and the perpetual motion machine is not necessarily that the former can be built and the second cannot—in fact, we do not know if it is possible to build a fusion reactor. The difference is epistemic and not ontological. The vision of the perpetual motion machine runs counter to accepted scientific knowledge, namely the laws of thermodynamics. Whether a vision is a fantasy, is a function of knowledge status and the thought collective that takes a stand on that knowledge status. There exist thought collectives that consider fusion reactors a fantasy (at least as practicable and safe sources of energy). Conversely, there are thought collectives outside mainstream science who consider perpetual motion machines as realistic visions.[4]

In the mentioned previous attempt (Strand, 2022) I argued that the imaginary of precision medicine as expressed both in biomedical science and policy, indeed is a fantasy because it runs counter to established knowledge. This includes scientific knowledge about the complex interrelationships between illness and disease, as well as about the complex interactions between the mind and body of individual patients and their ecology. I did not insist on the part of the Merriam Webster's definition that claimed that fantasies happen in response to psychological need; in fact, there may be many reasons for engaging in fantasy, including the explicit choice to take an oppositional or agnostic attitude to established knowledge. Given that fallibilism is a healthy philosophy of science, there is nothing intrinsically wrong in pursuing a research programme that contradicts what is held to be established knowledge. The problem is rather one of moral and social responsibility, above all the responsibility to take seriously the implications of the established knowledge that one wishes to dismiss or put into brackets. When fantasies are translated into projects and action programmes, they can be costly both in terms of investments and negative side effects. Sometimes, the side effects were not surprising at all from the perspective of the knowledge being bracketed or dismissed.

In what follows, I wish to pursue critique by means of yet another concept for what we could call inadequate or problematic practices of imagination, namely that of *escapism*. Again, I shall align with commonsense and apply the Merriam Webster

[3] See https://www.merriam-webster.com/dictionary/fantasy (accessed May 12, 2024).

[4] See for instance https://cosmicvault.tripod.com/page06.html (accessed May 12, 2024).

definition: Escapism is the "habitual diversion of the mind to purely imaginative activity or entertainment as an escape from reality or routine".[5] It may be seen as yet another turn of the screw away from adequate imagination: Not just producing unrealistic mental images, but in order to *escape from reality*. Again, the concept clearly is an epistemic one: In order to apply it, we need to take a stand on what is regarded as valid knowledge about reality. *Prima facie*, visions about fusion reactors or perpetual motion machines could both be very much grounded in concerns about reality, especially concerns about energy security. The same could be said about the visions of precision medicine and the real-life concern about, say, the suffering and premature death caused by cancer diseases and the hope to develop better treatments and cures. "Escapism", however, can also be seen to have a political dimension to it. Imagine a medical doctor placed in a university hospital in the middle of a war zone, surrounded by hundreds if not thousands of soldiers and civilians with life-threatening injuries. Let us imagine this poor doctor equipped with few or no means to alleviate the suffering of these soldiers and civilians. Perhaps he does have access to a research laboratory where he could pursue a molecular oncology project to better distinguish between genetic variants of prostate cancer in that context. Such a pursuit might definitely be interpreted by bystanders as escapism.

That type of bystander's interpretation—of identifying escapism within the war zone—is exactly what I intend to engage in. The war zone that I shall refer to, is the war that human civilisation seems to be mindlessly waging against nature, with casualties who are both human and non-human. The labels "Anthropocene" and "the Great Acceleration" will be used to describe this state of war. I shall interpret some of the contemporary imagination about the future of digital technologies as a way to try to escape from that stark reality. And I shall briefly discuss the normative questions of how we who at least believe that we are bystanders (to some extent) to that escapism, or at the very least who are still capable of stuttering some kind of critique of it, should relate to it. If we see digital escapism as a sign of a crisis of imagination, what could be our responses to that crisis? How can we move beyond it?

2 Anthropocene and the Great Acceleration

As clarified above, a claim about escapism has to be grounded in claims about reality, or at least claims about what should be considered as valid knowledge about reality. One strategy to put forth such claims, is to align with scientific consensus (or in the absence thereof, scientific establishment). While scientific research at its very forefront typically is directed to issues that are not settled, one can look for scientific consensus (or its proxies) in textbooks or in comprehensive reports from boundary organisations that have the mandate to communicate scientific knowledge to governmental institutions and policymakers. I shall do the latter. The unequivocal

[5] See https://www.merriam-webster.com/dictionary/escapism (accessed May 12, 2024).

message from such boundary organisations is that humanity's natural environment, the biosphere of our planet, is in rapid and dangerous deterioration. Let us spend a few moments on dwelling on their message.

The UNEP and the International Resource Panel issued their Global Resources Outlook 2024 report in March 2024 (UNEP, 2024). It states that "material use has increased more than three times over the last 50 years. It continues to grow by an average of more than 2.3% per year." (p. xiv) The world situation summarized as a *triple planetary crisis* (climate change, biodiversity loss and pollution).

The 2023 Synthesis Report of the IPCC states:

> Widespread and rapid changes in the atmosphere, ocean, cryosphere and biosphere have occurred. Human-caused climate change is already affecting many weather and climate extremes in every region across the globe. This has led to widespread adverse impacts and related losses and damages to nature and people (high confidence). Vulnerable communities who have historically contributed the least to current climate change are disproportionately affected (high confidence). (IPCC, 2023 [5])

The first Global Assessment Report on biodiversity was published by the IPBES in 2019, stating that "Nature and its vital contributions to people, which together embody biodiversity and ecosystem functions and services, are deteriorating worldwide" (IPBES, 2019 [10]).

The European Environment Agency, which is a governmental agency with the mandate to inform the European Union with scientific knowledge about the natural environment, stated the following in the executive summary of the 2020 edition of their flagship report "The European environment – state and outlook":

> In 2020, Europe faces environmental challenges of unprecedented scale and urgency. Although EU environment and climate policies have delivered substantial benefits over recent decades, Europe faces persistent problems in areas such as biodiversity loss, resource use, climate change impacts and environmental risks to health and well-being. Global megatrends such as demographic change are intensifying many environmental challenges, while rapid technological change brings new risks and uncertainties (EEA, 2019).

The report considers in detail the prospects of meeting EU policy objectives and targets on nature protection and conservation, the transformation into a low-carbon and resource-efficient economy and of protecting human health and well-being from environmental risks. The general conclusion, with some positive exceptions, is that Europe is not on track and that the objectives and targets are not going to be met by 2030 or 2050.

We could extend this analysis to indicators of human wellbeing and development, or to the progress on Sustainable Development Goals. The overall picture by time of writing—2024—is not only that progress is insufficient. It is a picture of multiple and mutually related environmental, economic, and humanitarian crises, the so-called "polycrisis". We are at a state in which official bodies, such as the

Think Tank of the European Parliament, have begun to use terms such as "permacrisis".[6]

From the perspective of environmental science, the multiplicity and chronicity of what appears as crises, is the consequence of human action over time, by which human influence on the natural environment has become so strong that it by far exceeds the biosphere's capacity to regenerate and restore, that is, it exceeds the so-called planetary boundaries. Moreover, the influence has been and still is accelerating, to the extent that it has been suggested that we have moved into a new geological epoch, the "Anthropocene", in which Earth-scale processes are notably driven by human activities (Crutzen, 2002). Will Steffen and coworkers have meticulously documented how a large set of socio-economic and earth system trends since 1750 show "hockey stick" or J-shaped curves, that is, with a pattern of rapid acceleration (Steffen et al., 2015). This has been called "The Great Acceleration".

While knowledge about their existence will remain uncertain (until it is too late), the research front of environmental science is in serious discussion about climate and biosphere tipping points, that is whether the planet's ecosystems can go into irreversible transitions that dramatically change (and most likely aggravate) the living conditions of humans and other species (Duffy et al., 2021; McKay et al., 2023; Wunderling et al., 2024).

None of this information is likely to be new to the reader of this chapter. The information has become commonplace. One might ask why it is relevant to mention all of this in a volume that sets out to discuss developments in ICTs and in particular the rapid development of the technologies labelled "AI", "artificial intelligence". However, the title of the volume is "Digital Environments & Human Ecology". Of course, we should discuss how AI is becoming part of the human house and the human ecology. However, two most basic ecological questions about a species are those of its identity and its survival. We are living at a time where the scientific establishment reports that we are in the middle of massive environmental deterioration that causes a mass extinction of species, and where the same scientific establishment is seriously considering the possibility that this development triggers tipping points that might even lead to human extinction. I cannot help but think that this fact has to be relevant to any conversation about human ecology.

3 Bios and Geos

Still, if escapism has to be defined against a backdrop of "reality", we should admit that there are many realities, in the sense that there are innumerous instances of subject positions and lived experience. For a single mother struggling to find food and shelter to her child and herself, reality is much more than the academic question

[6] See https://www.europarl.europa.eu/thinktank/en/document/EPRS_ATA(2023)747460 (accessed May 12, 2024).

of future ecological tipping points. What I am engaging in at this stage of the text is a language game for scientific and political elites, for privileged actors who can allow themselves to debate policy and governance and read this book chapter while taking today's food and shelter for granted. At the same time, this observation does not imply that such games are not real, serious or important. Indeed they are, potentially also for future food and shelter.

As for scientific and political elites, Jean-François Lyotard (in)famously noted a generation ago that they no longer could contain their own and others' imagination within one grand narrative of modernity and progress (Lyotard, 1984). 30 years later, Kjetil Rommetveit identified two major policy narratives in the Global North/West, namely those of *Bios* and *Geos* (Rommetveit et al., 2013). The Geos narrative is the one presented in the previous section, of climate change, environmental degradation, pollution and extinction. It is above all based in the system sciences—ecology, earth science, climate science and others. Its origins are old, at least going back to Thomas Malthus. It only came to its rise, however, in the 1970s, with Rachel Carson's *Silent Spring* in 1962, Kenneth Boulding's idea of Spaceship Earth, and the first photos of the Earth taken from space in the 1960s laying the ground. In its core, the Geos narrative states a fundamental concern for a finite and fragile planet and issues a strong warning about the limits to growth (Meadows et al., 1972). As witnessed in the previous section, the sciences informing this narrative have developed into an extensive and solid body of knowledge, and the concerns from the Geos narrative can be found in national, international and global policy, such as in the EU, the OECD and the various bodies of the UN. In terms of action, Geos calls for stopping and fundamentally changing current production and consumption practices and argues that our societies will remain unsustainable unless they go through deep and transformative change.

In parallel, however, the Bios narrative has also developed and consolidated in the world of science as well as that of policy. Bios is the narrative of innovation, competitiveness, job creation and economic growth. Just like Geos, it builds on sciences, but different sciences: It is informed by molecular biology, biotechnology, nanoscience, informatics and ICT development, robotics and above all neo-classical economics. It tells a tale of unprecedented prosperity and immense progress in terms of molecular understanding and the creation of new possibilities such as new materials, industrial processes and technologies. It bears the promise of unlimited growth. In terms of action, Bios calls for acceleration: We should continue to invest in basic (techno-)science and speed up rather than stop (Fig. 1).

Modern societies have written into their institutions that policies should be informed by scientific knowledge. The headache of policymakers in the last quarter of the twentieth century and into the 21st, is that both Bios and Geos are science-based while they are extremely difficult to reconcile. Bios comes with its own action programme that has remained much the same since the conception of Big Science around World War II. It is also well aligned with industry and commerce in capitalist economies of mass consumption. Geos, on the other hand, states that we should "stop" (as in "reduce emissions") and "change", but this advice is mainly negative, in the sense that it specifies what *should not* be done more than what *should* be

	BIOS	GEOS
Main symbol	 Double helix	 Gaia
Grand narrative	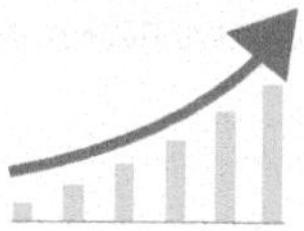 Growth	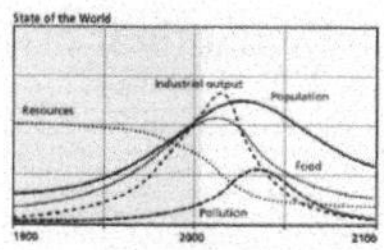 Limits
Main sciences and technologies	Evolutionary theory, molecular biology, ICTs, cognitive science, nanoscience, converging technologies, neo-classical economics, etc.	Earth system science, ecology, geosciences, climatology, Big History, ecological economics, etc.
Institutions	Nation state	Post-national

Fig. 1 Bios and Geos (Kovacic et al., 2019), reprinted as Creative Commons with license CC-NC-ND 4.0.

done. It needs further translation into active and positive characteristics. Moreover, it is in itself better aligned with idealistic civil society actors than with industry and commerce, that is, with non-profit rather than profit. In terms of policy, it can easily be implemented into conservation, protection, restriction and prohibition. To the extent that the Geos narrative is to be actualised in economic policy, however, it is in need of some kind of creative translation. Several such translations have been invented during the latter decades, including practices such as carbon trading and visions such as decoupling, the circular economy and, as we shall see in the next section, "the twin green and digital transition".

The idea of "decoupling" is central to the translations from Geos to Bios both in EU and global policy. The following excerpt from the opening paragraphs of the European Green Deal is characteristic:

> The atmosphere is warming and the climate is changing with each passing year. One million of the eight million species on the planet are at risk of being lost. Forests and oceans are being polluted and destroyed.
>
> The European Green Deal is a response to these challenges. It is a new growth strategy that aims to transform the EU into a fair and prosperous society, with a modern, resource-efficient and competitive economy where there are no net emissions of greenhouse gases in 2050 and where economic growth is decoupled from resource use. (EC, 2019)

The Geos narrative is unmistakably present in the first paragraph, while its concern is translated into Bios in the second paragraph.

There is a wealth of studies that support the claim that the vision of decoupling economic growth from resource use is a fantasy (EEA, 2021; Kovacic et al., 2018; Parrique et al., 2019). In particular, the idea that the circular economy can lead to a decoupling of growth and resource use, has been refuted theoretically as well as empirically (Giampietro & Funtowicz, 2020; Kovacic et al., 2019). Interestingly, some authors analyse its prominence with concepts similar to fantasy and escapism, namely as legend (Giampietro & Funtowicz, 2020) and delirium (Giampietro, 2018), in the sense of a dysfunctional denial of physical reality. Still, these visions continue to be prominent in public policy, and also supported by policymakers and scientists by what their critics characterize as cherry-picking of facts.

4 Digital Escapism in Public Policy

An even stronger candidate than the circular economy for the most escapist element in public policies in the Global North/West in the 2020s, however, is the promise of the "twin transition", understood as a coupled "green" and "digital" transition. I shall present an example of policy discourse that perhaps is somewhat extreme but still not atypical of what is found in policy documents and public statements especially in the EU. A key document in this respect is the EU's Strategic Foresight Report 2022, which was devoted specifically to the twin transition. While not in any way being the more panegyric document in this genre, consider this passage:

> Digital technologies could play a key role in achieving climate neutrality, reducing pollution, and restoring biodiversity. By measuring and controlling inputs, and with increased automation, technologies like robotics and the internet of things could improve resource efficiency and strengthen the flexibility of systems and networks. Energy efficient blockchain-based data management across the lifecycle and value chain of products and services could galvanise the progress towards a more circular economy and competitive sustainability. Digital technologies could also support monitoring, reporting and verification of greenhouse gas emissions for carbon pricing. Digital product passports enable enhanced material, component and end-to-end traceability and make data more accessible, which is essential for viable circular business models. Digital twins could facilitate innovation and the design of more sustainable processes, products, or buildings. Quantum computing will facilitate simulations too complex for classical computers. Space-based data technologies providing real-time global information monitor progress towards sustainability. Data sharing or gamification can increase public participation in steering the transitions and co-creation of innovations. (EC, 2022)

We may agree that all of this *could* happen, depending on how we qualify such terms as "key role", "improve", "support", "facilitate" and "increase". Taken at face value, the entire passage presents weak, uncommitted claims that ICTs *could* contribute towards sustainability, without taking a stand on whether it is likely to happen, or if these contributions would matter, given that what is called for—by science—to achieve sustainability, is deep, fundamental, transformative change of production and consumption patterns. And to be fair, the report also mentions the many possibilities that digitalization may be antagonistic to the green transition.

What happens, however, both in this very text above and its reception in the policy landscape, is a reification by which what "could" happen, gradually shifts to what *will, should* and *is going to* happen. By page 10, the Strategic Foresight Report states that "The twin transitions will result in profound shifts in the EU labour market and related skills." Within months after the publication of the report, in February 2023, Director-General Roberto Viola of DG CONNECT, the "ICT ministry" of the European Commission, called the twin transition "a driving force for a fair and prosperous future".[7] At the same time, the website of World Economic forum published statements such as:

> A twin transition approach recognizes that there is a huge and largely untapped opportunity for technology and data to drive sustainability goals. Rather than treating digital and sustainability in isolation, a twin transition strategy combines these critical functions to unlock huge benefits in terms of efficiency and productivity.[8]

My point here is not to claim that these visions necessarily are fantasies, in the sense of running counter to established knowledge. I suspect that such claims indeed could be made, but that is beyond the scope of this book chapter. Rather, I claim that they are escapisms, in the sense of purely imaginative activities as an escape from reality. On the whole, this policy discourse does not even try to substantiate the visions with solid knowledge. There are occasional academic references, but they are few. Rather, the main feature of the discourse has already been displayed, namely vagueness both in modality and content. For institutions that normally demand that their contractors and other interlocutors conform to the language of "SMART"—"specific, measurable, achievable, realistic and timely", their own policy discourse is conspicuously vague and uncommitted. In a world of polycrisis and permacrisis, these texts seem to be tailored to try to keep up hope within a Bios discourse, promising that ICT innovation will—or indeed, *could*—ensure that everything is going to be fine. Digital technologies *could* "play a key role in [...] restoring biodiversity", to quote the Strategic Foresight Report. How exactly that is thought to happen, in a reality where the mass extinction of species indeed is accelerating due to loss of wilderness, unsustainable land use and ruthless exploitation of natural resources, remains in the blue and perhaps it does not really matter. What matters is that the text presents consolation, hope, relief, and a justification for moving the attention away from the real problem as it anyway is going to (or at least *could*) be solved.

5 Digital Escapism in the Scientific-Technological Elite

The same discourse—of the twin transition—is also found within the world of science, technology and industry. Already President Eisenhower warned in his final address of the "danger that public policy could itself become the captive of a

[7] See https://www.europeanfiles.eu/digital/twin-transition-as-a-driving-force-for-a-fair-and-prosperous-future (accessed May 12, 2024).

[8] See https://www.weforum.org/agenda/2022/10/twin-transition-playbook-3-phases-to-accelerate-sustainable-digitization/ (accessed May 12, 2024).

scientific-technological elite" (Eisenhower, 1961). Who is held captive by whom in the triple helix, might be difficult to ascertain; perhaps it is better to consider the relationships as ones of mutual dependencies and coproduction. From the perspective of ICT industry, it is of course the opposite of escapism to lobby for policies that support ICT industry. Readers who work at public universities and other public institutions of science and higher education might confirm, however, how readily their employers adopt policy signals. When policymakers ask universities to jump for digital innovation, university leaders are often ready to jump. The same is the case for public research funding organisations, as illustrated by how swiftly e.g. Science Europe adopted the twin transition discourse.[9]

Another, more subtle form of digital escapism—and more generally, hi-tech escapism—is seen, however, in the choice of research foci as well as in the imagery deployed in self-presentation of STEM research. I prefer not to provide specific examples, as they would be somewhat arbitrary and could be perceived as attempts to single out individuals or specific projects. I would encourage readers, though, to reflect on their observations of how STEM research in the "digital" field—ICT development, robotics, big data-driven science within the field of biomedicine, et cetera—self-presents. Do you share my impression of a two-fold escapism in that (1) the imagery typically is one clean, controlled, technological environments, with few signs of the suffering of the Anthropocene, such as degraded or destroyed natural environments, pollution, shanty towns, draught, migration due to war and violence, and the many other sources of suffering not the least in the Global South?—and that (2) a great number of research projects and research fields primarily address the needs and concerns of the wealthy and privileged? In the field of medicine, this bias or rather injustice is called the "10/90 gap", signifying that approximately 10 percent of the resources for health research being spent globally, are devoted to the health problems responsible for 90% of the global disease burden[10]. I am not arguing that all researchers ought to engage in applied research that directly tries to solve grand challenges or the most immediate or acute social problems. I do not argue against basic research and fundamental, theoretical science, or against applied research that addresses problems of the rich. The claim I am making, is that if indeed it were the case that the digital revolution was on its way to meaningfully engage in the problems of the Anthropocene, there is a remarkable lack of visibility of those problems in scientific conferences, journals, and other arenas for self-presentation of this research. Either the problems are not present at all, or they are represented by imagery that echoes the escapism in policy world: by promises of clean and pleasant technological fixes.

[9] See https://scienceeurope.org/our-priorities/green-and-digital-transition/ (accessed May 12, 2024).

[10] See https://www.globalforumhealth.org/About/10-90-gap/ (accessed May 12, 2024).

6 Digital Escapism in Mass Consumption and Private Life

Finally, the above-mentioned dictionary definition of escapism as "habitual diversion of the mind to purely imaginative activity or entertainment as an escape from reality or routine" sound almost as it was made for the description of early twenty-first century use of internet among the privileged. Since smartphones became ubiquitous in the 2010s, life in private and public spaces has changed dramatically, at least across the North/West. To enter a means of public transportation in the 2020s in a European city is to witness a population of humans staring into each their smartphone device, many of them also wearing headphones, consuming internet entertainment from platforms such as TikTok, Instagram, Facebook, YouTube, and others. The same can be seen in parks, restaurants and other public spaces, even among people who apparently go out together, as groups or as pairs. One can see dogowners staring into their phones as they are walking their dogs, and parents doing the same while accompanying their children to the playground. It is remarkable how rapidly such behaviour has become widespread and accepted.

The International Telecommunication Union reported that 98% of European youth (15–25 years) use the internet, and 99% of high-income youth globally use the internet.[11] The extent of the use varies between years and countries and demographical strata, and I do not intend to go into the subtleties of this empirical field. However, one finding from the Pew Research Center is that about half of American teenagers in 2022 claimed to be on their smartphones "almost constantly".[12]

I have noted above how claims about escapism have to be founded in a claim about what constitutes reality. For children growing up in large parts of the world in the 2020s, the internet, social media and smart phones make up an important element of reality. In the context of the discussion that I try to raise in this book chapter, however, two claims of escapism seem relevant. First, on the level of the public sphere and in the interest of a well-informed political debate, there is a huge discrepancy between what the sciences of Geos report about the Anthropocene, the severe ecological risks we are confronted with and the need for change, on one side, and the entertainment-consuming apathy of large parts of the public, on the other. Indeed, in his last work, Bruno Latour hypothesized that the apathy grows with the direness of the scientific messages (Latour & Schultz, 2022). The scientific establishment of Geos has tended to assume that the discrepancy is due to an information deficit and that if only people were sufficiently warned, they would become engaged in the change towards sustainability. Unfortunately, this does not seem to be the case. It seems that people—and here I am not in any way excluding myself or the intellectual class or any other group for that matter—need to feel hope in order to engage. The unsolved challenge of the Geos discourse is to create or sustain some

[11] See https://www.itu.int/itu-d/reports/statistics/2022/11/24/ff22-youth-internet-use/ (accessed May 12, 2024).

[12] For details, consult https://www.pewresearch.org/internet/2022/08/10/teens-social-media-and-technology-2022/ (accessed May 12, 2024).

hope without falling into the false hopes of the prevailing translations into Bios (Kovacic et al., 2021). Catastrophism is not going to win sufficient hearts and minds.

Secondly, and here I take the risk of revealing myself as value conservative, to the extent that we take some traditional elements of human experience as fundamental aspects of reality, there are signs of digital escapism that are striking if not downright troubling. A key case is internet pornography and digital sex industry, which has developed from simple mass distribution of pornographic pictures and films on the internet, to new technological phenomena emerging in the twenty-first century, such as internet-mediated sex crimes, a growing sex robot industry, pornographic "deep fake", and the growth of AI-powered chat bots that simulate friendship and romantic relationships. Such troubling phenomena are rarely discussed in the clean, sterile and pleasant policy discourse about the digital transition. When discussing human ecology, however, it seems relevant to contemplate that ever more young men (and some women) find the idea of a "digital girlfriend" as attractive the idea of a real girlfriend. I propose that analytical philosophers take the challenge to do a proper analysis of such terms as "digital girlfriend". In the same way as Marcos and Bertolaso in this volume show why "artificial intelligence" is not intelligent, there is a need to show why a piece of software is not a girlfriend in a reasonable sense of the word. If we are not able to show that and argue convincingly why fully engaging with a real human being is infinitely richer and more valuable than with a computer programme, humanity indeed seems to be in danger.

7 Beyond Escapism: Staying with the Trouble and Finding Beauty in a Wounded World

In these times, what duties should we attend to as public intellectuals? What virtues should university academics of this and the next generation embrace and promote? I will argue for the importance of hope. As long as the Geos discourse is not seen as offering hope, we should expect ever more digital escapism. This is especially likely because, as was noted, one party's escapism is another party's business model and source of tremendous profit.

Respecting the scientific fact of limits to growth, however, false hopes should be rejected. The hope of a clean, swift, pleasant technical fix of the problems of the Anthropocene is contrary to knowledge; indeed, a fantasy. Hope has to be sustained in the presence of honesty about the fact that humanity may have to live for centuries in a world that experiences severe ecological disruption. In the words of Donna Haraway, we live in a wounded would. The challenge is not to escape into daydreams and learn to stay with the trouble (Haraway, 2016).

To the extent that hope has to be connected to an idea of progress, it would be seen to lie in other domains than those of material consumption, wealth or power. From the facts of biophysical imbalances between man and planet, it follows that the cultural, professional and human development to be called for, should have wider scope than the material one. We should protest against the equivalence between growth and economic growth. Growth can also be personal, civilisational, moral and spiritual (EEA, 2021).

Haraway is among those who warned against an unfortunate connotation of the term "Anthropocene". The choice of the word was grounded in the observation that human activity has reached the level where it acts as a driver of geological change. The unfortunate connotation is that it sounds as if we are the only important life form on the planet. Going back to the ecological philosophy of Arne Næss, however, we should insist that humans neither should nor can live alone, without other species, for a multitude of reasons (Næss, 2005; Oliver, 2020). Human flourishing and multispecies flourishing are in this view two sides of the same coin.

This point may sound abstract but is highly concrete. Every day, we have the choice to either continue to stare into the smart phone or look up to notice and connect with other living beings, human and non-human. Hope, beauty and agency can be sought in the simple act of looking up, smelling the flowers and listen to the birds (Strand & Kovacic, 2022). "Staying with the trouble" may sound hard, but in this sense it is also the quest of finding, showing and walking avenues to connectedness, love and compassion with the humans and other living beings of our beautiful planet. This quest is one of action and imagination, and I believe, the one that could take us further on the way out of the crisis of imagination. Recalling that information and communication technologies are technologies, the question would then not so much be whether they can be described as part of the human environment, but to put our efforts into imagining how ICTs can be employed along such avenues, in service to the real living beings of the real world.

Acknowledgments I would like to thank Marta Bertolaso and the Campus Bio-Medico University in Rome for their collaboration and support, and Ingeborg Rønning and Adriaan Eeckels for inspiring conversations during the preparation of this manuscript. This work was partially funded by the Research Council of Norway through its Centres of Excellence funding scheme (CCBIO, project number 223250).

References

Crutzen, P. J. (2002). Geology of mankind. *Nature, 415*(6867), 23–23. https://doi.org/10.1038/415023a

Deleuze, G. (1994). He Stuttered. In *Gilles Deleuze and the theater of philosophy* (pp. 107–114). Routledge.

Duffy, K. A., Schwalm, C. R., Arcus, V. L., Koch, G. W., Liang, L. L., & Schipper, L. A. (2021). How close are we to the temperature tipping point of the terrestrial biosphere? *Science Advances, 7*(3), eaay1052. https://doi.org/10.1126/sciadv.aay1052

EC. (2019, December 11). *Communication from the commission to the European parliament, the European council, the council, the European economic and social committee and the committee of the regions the European green Deal, COM/2019/640 final* [website]. Publications Office of the EU; Publications Office of the European Union. Retrieved May 12, 2024, from https://

op.europa.eu/en/publication-detail/-/publication/b828d165-1c22-11ea-8c1f-01aa75ed71a1/language-en/format-PDF/source-288831174

EC. (2022). *Communication from the commission to the european parliament and the council. 2022* Strategic foresight report: Twinning the green and digital transitions in the new geopolitical context. Publications Office of the European Union. Retrieved May 12, 2024, from https://eur-lex.europa.eu/legal-content/EN/TXT/?uri=CELEX%3A52022DC0289&qid=1658824364827

EEA. (2019). *The European environment: State and outlook 2020: Knowledge for transition to a sustainable Europe*. Publications Office of the European Union.

EEA. (2021). Growth without economic growth. *Publications Office*. https://doi.org/10.2800/492717

Eisenhower, D. D. (1961). President Dwight D. Eisenhower's farewell address. National Archives. Retrieved May 12, 2024, from https://www.archives.gov/milestone-documents/president-dwight-d-eisenhowers-farewell-address

Giampietro, M. (2018). Anticipation in Agriculture. In R. Poli (Ed.), *Handbook of anticipation: Theoretical and applied aspects of the use of future in decision making* (pp. 1–35). Springer International Publishing. https://doi.org/10.1007/978-3-319-31737-3_23-1

Giampietro, M., & Funtowicz, S. O. (2020). From elite folk science to the policy legend of the circular economy. *Environmental Science & Policy, 109*, 64–72. https://doi.org/10.1016/j.envsci.2020.04.012

Haraway, D. J. (2016). *Staying with the trouble: Making kin in the Chthulucene*. Duke University Press.

IPBES. (2019). Summary for policymakers of the global assessment report on biodiversity and ecosystem services. *Zenodo*. https://doi.org/10.5281/zenodo.3553579

IPCC. (2023). *AR6 synthesis report: Climate change 2023*. Intergovernmental panel on climate change. Retrieved May 12, 2024, from https://www.ipcc.ch/report/ar6/syr/

Jasanoff, S. (2019). *Can science make sense of life?* (1st ed.). Polity Press.

Jasanoff, S., & Kim, S.-H. (Eds.). (2015). *Dreamscapes of modernity: Sociotechnical imaginaries and the fabrication of power*. University of Chicago Press. Retrieved May 12, 2024, from https://press.uchicago.edu/ucp/books/book/chicago/D/bo20836025.html

Kovacic, Z., Spanò, M., Piano, S. L., & Sorman, A. H. (2018). Finance, energy and the decoupling: An empirical study. *Journal of Evolutionary Economics, 28*(3), 565–590. https://doi.org/10.1007/s00191-017-0514-8

Kovacic, Z., Strand, R., & Völker, T. (2019). The circular economy in Europe: Critical perspectives on policies and imaginaries.

Kovacic, Z., Benini, L., Jesus, A., Strand, R., & Funtowicz, S. (2021). When the unspeakable is no longer taboo. *Issues in Science and Technology, 37*(4), 16–18.

Latour, B., & Schultz, N. (2022). *On the emergence of an ecological class: A memo* (J. Rose, Trans.; 1st ed.). Polity Press.

Lyotard, J.-F. (1984). *The postmodern condition: A report on knowledge* (G. Bennington & B. Massumi, Trans.; 1st Ed.). University Of Minnesota Press.

McKay, D. I. A., Sakschewski, B., Roman-Cuesta, R. M., Dakos, V., Flores, B. M., Hessen, D. O., Hirota, M., Kéfi, S., Obura, D., Reyer, C. P. O., Staver, A. C., Thom, D., Abis, B., Amaral, C. do Andersen, T., Bathiany, S., Beaugrand, G., Blenckner, T., Brovkin, V., … Wiltshire, A. J. (2023). Tipping points in the biosphere. In Global tipping points: Report 2023 (pp. 78–121). University of Exeter. Retrieved May 12, 2024, from https://research.wur.nl/en/publications/tipping-points-in-the-biosphere

Meadows, D. H., Meadows, D. L., Randers, J., & Behrens, W. W. (1972). *The limits to growth–Donella H. Meadows*. Potomac Associates. Retrieved May 12, 2024, from https://collections.dartmouth.edu/ebooks/meadows-limits-1972.html#epubcfi(/6/2[front_cover]!/4/1:0).

Næss, A. (2005). Self-realization: An ecological approach to being in the world. In A. Drengson (Ed.), *The selected works of Arne Naess* (pp. 2781–2797). Springer. https://doi.org/10.1007/978-1-4020-4519-6_128

Oliver, T. (2020). *The self delusion: The surprising science of how we are connected and why that matters*. Weidenfeld & Nicolson.

Parrique, T., Barth, J., Briens, F., Spangenberg, J., & Kraus-Polk, A. (2019). Decoupling debunked. *Evidence and arguments against green growth as a sole strategy for sustainability. A study edited by the European Environment Bureau EEB.*

Rommetveit, K., Strand, R., Fjelland, R., Funtowicz, S., & Saltelli, A. (2013). *What can history teach us about the prospects of a European research area*. European Commission Joint Research Centre.

Steffen, W., Broadgate, W., Deutsch, L., Gaffney, O., & Ludwig, C. (2015). The trajectory of the Anthropocene: The great acceleration. *The Anthropocene Review, 2*(1), 81–98. https://doi.org/10.1177/2053019614564785

Strand, R. (2022). The impact of a fantasy. In *Personalized medicine in the making: Philosophical perspectives from biology to healthcare* (pp. 311–324). Springer International Publishing.

Strand, R., & Kovacic, Z. (2022). The data of the rose. In M. Bertolaso, L. Capone, & C. Rodríguez-Lluesma (Eds.), *Digital humanism: A human-centric approach to digital technologies* (pp. 49–57). Springer International Publishing. https://doi.org/10.1007/978-3-030-97054-3_4

UNEP. (2024). *Global resources outlook 2024*. Retrieved May 12, 2024, from http://www.unep.org/resources/Global-Resource-Outlook-2024

Wunderling, N., von der Heydt, A. S., Aksenov, Y., Barker, S., Bastiaansen, R., Brovkin, V., Brunetti, M., Couplet, V., Kleinen, T., Lear, C. H., Lohmann, J., Roman-Cuesta, R. M., Sinet, S., Swingedouw, D., Winkelmann, R., Anand, P., Barichivich, J., Bathiany, S., Baudena, M., et al. (2024). Climate tipping point interactions and cascades: A review. *Earth System Dynamics, 15*(1), 41–74. https://doi.org/10.5194/esd-15-41-2024

Technological Humanism & Sustainable A.I.

Alfredo Marcos and Marta Bertolaso

Abstract In this chapter we first disentangle the philosophical categories that are at stake when talking about 'digital environments' having in mind the question about sustainable AI. As ethics follows ontology, in the second part, we continue addressing some ontological questions on artificial intelligence (AI). The most elementary is whether the AI exists. We will argue that what actually exists is not AI as such but *AI systems* that emerge from the combination and dynamic interactions of human beings and machines. In this sense also environments that are digitally mediated deserve attention, philosophical and scientific analysis. In other words, we must go from considering AI systems as technical systems with social consequences to considering them as technically implemented social systems. People are part of the AI systems, as designers and users. As a consequence, we can say that in AI systems the intelligent part is not artificial, and the artificial part is not intelligent. The question about sustainable AI is thus not a technical problem, but an ontological and anthropological one, a problem of human ecology. When we accept that intelligence can be found in a simple device, we succumb to what might be called the *Toy Story effect*. Toys do not play by themselves, as well as machines do not have intelligence. Both playing and understanding require the concurrence of a human being, and this fact has a number of ethical consequences. Given these reflections, we conclude that considering the nature and the virtues that human beings can develop in environments that are also digitally mediated, opens new scenarios for education, policies and innovation that has to be sustainable by design.

A. Marcos (✉)
University of Valladolid, Valladolid, Spain
e-mail: amarcos@uva.es

M. Bertolaso
Università Campus Bio-Medico di Roma, Rome, Italy
e-mail: m.bertolaso@unicampus.it

A. Fabris, S. Belardinelli (eds.), *Digital Environments and Human Relations*, Human Perspectives in Health Sciences and Technology 150,
https://doi.org/10.1007/978-3-031-76961-0_10

Keywords AI · Delegated control systems (DeCo systems) · Information · Intentionality · Simulation · Function · Human ecology · Sustainability · Environments · Virtues

1 Introduction

Today we live in multiple environments. Naively, this notion denotes multiple spaces and contexts, such as, for example, the space inhabited by a person or the space that, generically, surrounds a thing but also, more broadly, the physical-chemical and biological conditions in which all living being live. In a figurative sense, the term can also represent the social and cultural dynamics that determine personal growth; in this second meaning of the term, a conceptual reference to the human being and his agency dimension in the world is still evident.

The notion of environment, therefore, semantically refers to multiple nuances and uses. Moreover, if we consider the technological development, we have to acknowledge that we live in environments that are more and more digitalized. Such transitions are bringing with them a new language such as the use of the term 'digital environments'. How should we understand such emerging concept? Should 'digital environments' be considered an additional 'environment' at the same level and with the same nature of the natural ones or should we deepen what is really at stake in the process of adopting such language to better understand the emergent risks and possibilities of the emergent technologies for an equilibrated human flourishing in the so called 'digital environments'? What epistemological and ontological issues are at stake? How can we reflect on them for an adequate topology of the environment notions and of the personal and social responsibilities we have to encourage and foster?

In order to answer these questions, in this chapter we first disentangle some philosophical categories that are at stake when talking about 'digital environments'. Secondly, as ethics follows ontology, we continue addressing some ontological questions on artificial intelligence (AI). We will argue that AI does not exist as such; what does exist, instead, are the *AI systems* that emerge from the combination and dynamic interactions of human beings and machines. In other words, we must go from considering AI systems as technical systems with social consequences to considering them as technically implemented social systems. People are part of the AI systems, as designers or users.

In the third part, therefore, the question about sustainable AI poses not a technical problem, but an ontological and epistemological one. Part of the argument follows also the consideration that when we accept that intelligence can be found in a simple device, we succumb to what might be called the *Toy Story effect*. Toys do not play by themselves, as well as machines do not have intelligence. Both playing and understanding require the concurrence of a human being, and this fact has a number of ethical consequences. Something similar happens with the notion of metaverse. Therefore, we conclude that human beings' flourishing and virtues have to be developed in also digitally mediated environments, and that this requires new awareness and it opens new scenarios for education, policies and innovation that have to be sustainable by design.

2 Environments and Human Living Beings: An Integral Ecological Approach

Philosophically, we can start from the incontrovertible fact that the notion of environment is a concept full of meaning that refers to the age-old question of the relationship between context and living being and the mutual influence and relationship between the two. Different approaches can be adopted when trying to scientifically address this issue. Following Jakob von Uexküll and the notion of environment (*Umwelt*) he developed in his biological studies and in his book entitled "Animal Environments and Human Environments" (Uexkull, 2010; firstly published in German in 1933), it is to some extent commonly now acknowledged that the living beings might be considered as a zero point of a world or environment which is not unique and universal but constituted by and for the subject itself. The environment, therefore, is given to the subject in a changing and reciprocal exchange. The use of the plural—environments—is consistent with this idea. What also follows is that the environment cannot be enquired in isolation and that a physiological investigation that considers any living being as an object of research in itself, located in a unique and univocal world and that can be investigated in a universal way through a mechanistic approach, does not hold. At least, such reductionist approach is not useful when the question is about the sustainable development of a given living (eco)system.

In this sense, the theoretical revolution exemplified by Jakob von Uexküll mainly concerns the abandonment of the mechanistic perspective and the consequent reductionism represented by it in the ecological sciences. In doing so, his approach has contributed to dismantling the idea of the clock, made of mechanisms and gears, as the paradigmatic symbol par excellence of the constitution of the world and of the human being himself. Many works followed which recognized the notion of system and environment as constitutively relevant, undermining the modern reductionist views.

When coming to current digital transitions, the panorama seems to be more complex. If it is true that we live in physically characterized environments, we should ask the question whether the pervasiveness of the new digital tools is actually bringing us into a new kind of environments—"digital environment"—or what is actually at stake in these transitions in philosophical terms. We surf online from morning to night, we are always hyper-connected through social media, emails and instant messaging apps which have revolutionized communication. Chat-GPT is not an exception. We have the possibility now to develop economic transaction and market activities in contexts that are almost completely digitally mediated. Compared to the environments we were used to live, the digital imposes itself with a disruptive force, in the creation of a non-physical but nevertheless immersive environment that can modify our experiences. Artificial Intelligence (from now AI) is the technology behind all these scenarios, it amplifies our possibilities of choice, of knowledge organization, accelerates our decision-making processes while orienting them in many cases. In order to analyse this new way of living, we will delve into three ideas: the ecological approach to the environment, the notion of environment-world and the revolution imposed by the digital transitions.

2.1 "In-Being" as a Descriptive Character of the Notion of Environment

From the description of the various environments made above, it emerges that the notion of environment thematically refers to the notion of living which, in turn, necessarily opens up to the methodological question of how we live in the environment. Parallel to the conceptual development of the scientific notion of environment, which we historically date back to the last century, the problem of finding a description capable of emphasizing the relationship between a human living being and the surrounding environment emerged.

On the basis of the made considerations, the relationship between subject and environment can be usefully described according to the Heideggerian philosophical category of "in-being" (Heidegger, 2008). This expression, taken from the existential analytics of the German phenomenologist, seems to capture the vital and relational nuance that emerges from the common conception of the environment as a place of experience that the subject inhabits. In fact, the notion of "in-being" (*Da-sein*) does not have the objective of describing the physically connoted character of an object of being inside something, as in the case of the water that is found in the bottle or the wardrobe that is in a room. On the contrary, such category underlines the existential relationship that is established with the world, that is, inhabiting the surrounding environment according to the character of familiarity. In the text *Fundamental Concepts of Metaphysics* (Heidegger, 2001), Heidegger deepens and further refines the distinction between world and environment by describing a human world-environment in which the human being, by nature, structures the surrounding environment as a world, making it not only the place of action and experience but forging and transforming it as a historical-cultural space. Understood in this way, the environment of the human being is an open world-environment, flexible and indeterminate in principle but determinable through the action of the human being himself. The transformative action of the human being, therefore, produces an always mediated world. Technology is a natural and constitutive part of such mediation. As we have many times argued, human beings are in fact naturally technical beings, i.e. they dwell in their environments and change or adapt it to their needs and expectations through the *techné*, a notion embedded in the modern technologies too. In this sense, we consider the environment as 'natural' (whether natural, i.e. physical, or artificial). Artifacts are natural constituents of the human environment. But it seems clear that the digital mediation of some aspects of human environments poses new questions, as they change the perception and the possibility of action that human beings have in a given context. Considering the reciprocal interactions between the subject and the environments, this has to have consequences for the perception of the *Self* too.

2.2 Technological Mediality

Keeping in mind our initial question about sustainable AI, let's disentangle this point by considering that *mediation* is the fundamental fact of the technique and, therefore, of the technology with which the human being inhabits the world-environment and at the same time the characteristic that brings the concept of environment back to the centre of the contemporary debate. The advent of digital technologies has actually increased such level of mediality in human life and intersect previously isolated media fields, challenging the human "in-being" in a given environment.

What does this mean? As we discussed in another publication (Capone et al., 2023) when considering the impact of AI in the current digital transitions we have to move beyond a mere instrumental conception of the media, that is, thinking that "the purpose of digital mediation is merely the solution of a communication problem: representing any kind of content, conveying messages at high speed, at a minimal expenditure of information and at low cost (Shannon & Weaver, 1963)" (quoted in *ibidem*). The digital cannot be considered as a mere encoding as it does not merely deal with *the way* people solve a mathematical problem, listen to music or go shopping. "Media structure a model of relationship with things and these relationships are bearers of peculiar kinds of agency" (*ibidem*). Therefore, the human world-environment progressively takes shape as a mixture of different relationships, variously connected, each subjected to different mediations and liable to come into contact with certain media and not with others. Moreover, the media (be they platforms, particular technologies, information exchange protocols, methods of accessing data, etc.) direct human practices in certain directions and shape relationships within these practices.

2.3 Human Ecology and Technological Sustainability

According to Lady Bird Johnson's definition "*The environment is where we all meet, where we all have a mutual interest; it is the one thing all of us share* " (Johnson, 1967). This concept is the basis of ecological philosophy, which examines humanity through an eco-ontological lens, highlighting relationships and sharing (Marchesini, 2002). This perspective recognizes that everything in the world is interconnected, living beings and the environment are intrinsically linked, interdependent. Humanity is an integral part of the ecosystem, it is not separated from it, and every human action, every economic or social process has an impact on the environment and on human beings themselves. The environment is one of the fundamental dimensions of this interconnection, influencing and being influenced by every aspect of human life and the natural world. The above-mentioned environments—also technologically mediated—do not exist in the vacuum but have consequences for all the other living beings' environments and vice versa. It therefore makes sense to think of these issues in terms of *Human Ecology* to remind us that reflection on human behavior cannot be separated from the awareness of the interconnections that exist between us and the world around us. Somehow the relationship we have with ourselves, the care we are able to take of ourselves and others necessarily reflects our relationship with the world and the world we want to build and inhabit.

Philosophers such as Aldo Leopold and Arne Naess have developed innovative ethical paradigms that redefine human-environment interaction on the basis of a notion of environment that is fundamentally linked to that of resource: that is, the resources that the environment offers and all living beings they have an intrinsic value, and this must push humanity to reconsider its role in the Earth's ecosystem and to develop a deep sense of responsibility towards the environment in which it lives and operates (Leopold, 2020; Naess, 2016). The environment, therefore, as a constitutive dimension of the dynamics of a system, requires people to have an ethically responsible approach that is aware of the ecosystem value of the environment. From this come the objectives of preserving natural resources, which would continue to support life on Earth, and that of promoting long-term global well-being without compromising the ability of future generations to meet their own needs—sustainable development [see Brundtland Report (UN, 1987)]. A healthy environment is essential to ensure that this equation is met, and is also capable of adapting to change, which is crucial for the resilience of today's rapidly evolving world. The global challenges posed by the goals for the sustainable development (SDGs) follow this logic.

Technology contributes to these processes. In particular, digital technologies can be integrated into physical environments to create advanced sensors that allow interaction between humans and the surrounding environment, providing real-time data (monitoring), which can be made accessible through applications or digital interfaces (implementation). In this sense, the digital technologies can also make environmental 'resources' more transparent and allow us to influence them like never before. Network logics and the dematerialisation of digitally mediated relationships allow us to manipulate and orient environmental dynamics by adapting them to individual (personalization) and collective (interactivity) needs, allowing users to examine specific aspects, such as (i) identify complex relationships between the individual and various other factors, (ii) formulate predictions using systemic logics and (iii) make more informed decisions about the environment and people's health too. In both cases, nevertheless, digital technologies create interfaces that are nevertheless *representation of the environment-world*. Although such representations sometimes also end up being part of our environment, their representational nature what actually affects are the human relationships and their perception.

As in the case of technologies applied to the human body, also in this case we need to question ourselves about the constraints that hold a flourishing use of such technologies in mediating a perceived environment. As Hans Jonas said, losing the human for enhancement technologies would cause us to lose the axiological reference that allows us to evaluate its goodness or illicitness (also discussed in (Bertolaso & Marcos, 2023)). It thus makes sense asking the question of which axiological reference is necessary consider so as not to disorientate ourselves in incorporating technology into the environment around us. In this case we will talk about eco-technological sustainability, which deals with the still only human capability of making decisions. Beyond our possibility of action human intelligence is up to this: questions about meaning, and actions that follow a judgment. Values and virtues mediate this dual processes. A relational philosophical understanding of the human beings holds the possible integration of all these aspects.

3 What Is A.I.

After having contextualized the environment issues and highlights reasons for an integration of the philosophical questions at stake across the notions of environment, digital technologies and sustainability, the need for an ontological reflection emerges. As ethics follows ontology, the question about A.I. (what is AI) still needs to be addressed in the light of a philosophical account of the human beings as relational and technological beings, beings that are able to change the environment for their own sake and objectives.

Heidegger "in-being" notion requires, therefore, not only a deeper reflection about the human beings, but also about digital mediated environments and AI more in general.

Although we all have been reading and writing a lot about AI, we suggest now to consider AI ethical and political problems from a different perspective. Our ethical tribulations, in fact, almost always refer to deeper ontological problems that sometimes we have undermined. And so is the case regarding AI. Without addressing the ontological questions, the ethical debate becomes imprecise and superficial. Disentangling the ontological questions that are behind the ethics of the so-called AI, we find that the most elementary question we can ask in ontological terms is whether or not AI exists. Luc Julia (Julia, 2019), who led the SIRI development team at Apple, provocatively titles his book: *L'intelligence artificielle n'existe pas*. For his part, Erik J. Larson (Larson, 2021), in a similar vein, has published the book entitled *The Myth of Artificial Intelligence. Why Computers Can't Think the Way We Do*. In the advance of this book we can read: "We aren't really on the path to developing intelligent machines. In fact, we don't even know where that path might be

[…] AI will continue to improve at narrow tasks, but if we want to make real progress, we will need to start by more fully appreciating the only true intelligence we know—our own".

3.1 An Inappropriate Name

What do we mean when we say that there is no AI? Maybe it's (again) a naming problem. Perhaps there is something which we call AI, but for which we are using an inappropriate name, since it does not contain true intelligence or the intelligence it contains does not reside in the artificial part of this entity. It would, therefore, be necessary to propose a better denomination, which would not lead to confusion. "The term artificial intelligence – recalls Katharina Zweig, from the Algorithmic Accountability Lab at the University of Kaiserslautern – arose in the 1950s, when scientists wanted to raise money for their research. They thought it sounded like something the State would be happy to encourage. And now we hang on this name. Most computer scientists find it inappropriate" (Von Hopffgarten, 2021).

Machine learning or *deep learning* are equally confusing names, especially if we have to understand that it is the machine itself that learns. All these denominations—lures, it should be said—have a commercial, advertising, even propaganda function, but they do not respond to the truth of the thing. They immediately resonate with science fiction and the media headlines. Therefore, dreams and future scope terrors begin to thrive. However, no machine understands, or knows, or learns, or is capable of counting to two. People do it, with the help, sometimes, of machines. For this reason, other terms have been proposed, such as *assisted intelligence*, *expanded intelligence*, *human-centered artificial intelligence*, *decision support tools*… These names are more appropriate, since they indicate that the intelligent subject is a person, while the machine can assist or expand the intelligence of said subject. We could also talk, and I think it would be the most appropriate option, of ***Delegated Control systems*** (which we can abbreviate as *DeCo*).

The change in perspective that this name introduce could be summed up in a few words: we must move from considering AI systems as "technical systems with social [external] consequences" to considering them as "technically implemented social systems" (Hirschheim et al., 1995 [1]). In other words, people are also part of the AI systems, as designers, owners, maintainers, users, supervisors, lawmakers… It is in these people, and not in the artificial part, where the intelligence of these systems resides. In AI systems, the intelligent part is not artificial, and the artificial part is not intelligent. Machines cannot be intelligent by themselves, outside a human environment. This limitation does not respond to a technical problem that can be technically corrected, but to an ontological difference.

3.2 So, What Is Intelligence?

Now let we take a step back and reflect for a moment on the very concept of intelligence. This way, we will better estimate if certain entities deserve or not to be called intelligent. Or, more precisely, where intelligence resides in *a technically implemented social system*. Dictionary definitions of the word "intelligence" often refer to the ability to understand. It is also common for them to allude to the ability to solve problems. For instance, The Oxford Learner's Dictionary defines "intelligence" as "the ability to learn, understand and think in a logical way about things". For the *Diccionario de la lengua Española (Spanish Dictionary)*, intelligence is the "ability to understand or comprehend… [and, in a second meaning] The ability to solve problems". And something similar appears in the dictionary of the Italian language *Treccani*.

We know that the artificial part of an AI system is incapable of understanding by itself, without humans. You cannot even properly say that a machine counts or computes. Counting implies joining two (or more) moments (or things) and keeping them together in a single and identical conscious representation, understanding at the same time the similarity—not identity—and the difference between them. All this far exceeds the capabilities of a machine.

It is true, on the other hand, that AI can help us solve multiple problems (computing, writing texts, drawing, geolocation, logistics, telephone assistance, medical diagnosis assistance, advertising and a long etc.). But these problems are not for the artificial part of the system, but for the human designer or user of the system. For a facial recognition machine, recognizing or not recognizing a criminal is not a problem. It is a problem for people's safety, and the system can help us deal with it. Of course, the same system can be used to control the population of a country and to facilitate political repression there. Again, this is not a problem for the cameras or for the software involved. It is, undoubtedly, a problem for the human subjects of the country in question. Only a living being can suffer and die; only a person can wonder about the meaning of her life. Those are problems. And both a hammer or an abacus, and a computer network too, each in its own way, can help us deal with these problems (or make them worse). But this does not make these tools intelligent. We understand. We have problems. Not the machines.

Let's take a look from another angle. Sometimes the so-called AI is characterized by its simulation capacity. It simulates functions of human intelligence, it is said. However, simulating intelligence is not the same as being intelligent. Furthermore, the simulation only appears as such for the human being who observes it, not for the machine. The machine does not know that it is simulating intelligent behaviour. On the other hand, the very notion of function inexorably refers to a being for which a given effect is functional. Outside the human framework, the lights that come on and off on a screen, or the movements of a robot are mere effects. They do not fulfil functions. It is the human point of view that changes its ontology transforming mere effects into functions.

3.3 Back to the Ontology

So, the question is basically—let us insist—of ontological nature. Artefacts, in the Aristotelian tradition, in the Aristotelian ontology, are substances only in an accidental sense, i.e. by analogy. Living beings, and especially human beings, are substances in their proper and paradigmatic sense. As it is an ontological difference, the hope (or threat) of annulling it through technological sophistication is illusory, a mere category mistake.

What has been said so far affects any technological system (washing, transport, energy production, communication…). All of them, if they are placed outside of the human environment, lose their functionality, they become plain systems of physical effects. Since their ontology is given by their function, they also lose their ontological rank, they stop being what they were. A washing machine placed on Neptune is no longer a washing machine. Nevertheless, the ontology of the so-called AI systems depends even more intensely on the human gaze, since they are located in the realm of the intentional, that is, of the semiotic. In this area, the entities are sustained on three supports. If we remove one of them, they will collapse, as happens to the stools. Charles S. Peirce makes it clear: "All dynamical action, or action of brute force […] takes place between two subjects […] But by semiosis I mean, on the contrary, an action or influence which is or involves a cooperation of three subjects, such as a *sign*, its *object* and its *interpretant*, this three-relative influence not being in any way resolvable into actions between pairs" (Peirce, 1935 [484]).

What do we mean when we say that a machine stores or processes my financial or medical data? We say that a certain electromagnetic state of the machine (*sign*) is related to my payroll or my blood pressure (*object*). Obviously, there is no physical relationship between them, but rather a semiotic relationship that is established through a person (*interpretant*) capable of understanding or interpret—with the help of certain interfaces—electromagnetic states as income or blood pressure. Similarly, the machine only plays chess or *go* if a person can relate the physical states of the machine to these games traditionally played by humans. The case of chess is very illustrative: when a machine finally meets certain expectations, it is at the same time deprived of the mythical aura that surrounded it when it was just a project; it is reduced to the level of the prosaic, devoid of ghost and glamour. Look at the poor *Deep Blue*, who knew glory days, raising now museum dust.

It is exactly the same in the case of the now famous ChatGPT. Its performances are really impressive. It is very helpful in multiple tasks. But, in spite of its astonishing attainments, no one thinks that this software understand anything. Rather, it has become clear to us that great linguistic achievements can be attained without understanding anything at all (as John Searle advanced years ago).

So, we've all learned, at the end of the day, that a chess-playing robot, or a chatbot, is about as interesting as a vacuum-cleaning robot. Without an *interpretant*, the machine just changes from one physical state to another. It is no longer part of an intelligent system. It is just a piece of matter, like a washing machine on Neptune. Why is it so hard for us to accept it?

4 The *Toy Story* Effect

We tend to imagine that in our absence the artefacts continue to have the same entity as in our presence. Thus, we imagine that a machine that is part of an AI system, together with certain people, continues to be intelligent even if it does without the gaze of those people. But this mirage is not due to an excess, but to a lack of imagination. It is not easy to imagine what the world looks like when the world is not seen by anyone. The look of the human being sustains the being of the artificial. Without the gaze of a person, the artificial is flattened; it becomes in pure physical reality. Hence the difficulty of imagining. It's easier to dream that everything stays the same when I stop looking. It is what we could call *the Toy Story effect*. The child's hand and eye turn a piece of green plastic into a shy dinosaur. The child imagines that when he leaves the room, the dinosaur is still there. He can't think of it like the inert piece of green plastic it is when the child himself leaves the playroom or falls asleep.

What do we imagine would happen if the human beings left the room (Chinese?), if they were left out of the AI systems? For some, this will occur from the point they call *singularity*. From there the machines would generate other smarter machines, a post-human world controlled by robots. But perhaps we could imagine, on the contrary, that machines left to themselves would soon fail by virtue of the general tendency towards entropy, design and construction defects, as well as the difficulty in obtaining stable energy sources; they would decay and be re-incorporated into the natural world, into the physical processes (such as erosion) and chemical processes (such as oxidation). Even into the biological processes: the most probable post-human landscape is not that of the Earth governed by intelligent robots, but that of a leafy jungle that hides in its bowels, along with the stones of ancient temples, authentic filth of silicon, plastic and metal. In fact, every machine has to be maintained, that is, led by the hand by people ("hand" in Latin "manus", and from there, in English "to maintain"). Every AI system requires maintenance. And the more sophisticated ones require more maintenance, not less.

In short, data are data about something, intelligence is intelligence about something, and also the information. They are triadic, semiotic, intentional entities. The electromagnetic (or quantum) states of a computer are not data by themselves, unless an *interpretant* manages to connect it with an *object*. (And this dependence of the data on a consciousness is extreme in the case of the so-called *synthetic data*, whose relationship with the truth is so difficult to establish). Without people, an AI system immediately ceases to be intelligent. There are no more data. It no longer understands anything. It no longer simulates anything. It no longer serves any function. Its troubles are finally over. What we called information is diluted. All decision ceases to be such.

In order to avoid misunderstandings, we should change "AI" to a less tortuous name, since the disturbance generated by a bad name ends up being projected onto anthropology itself. Thus, the image of human intelligence is degraded or reduced to an algorithmic game. Such a game could be played on this or that material support (carbon, silicon… it doesn't matter). So, the human being himself comes to be understood in dualistic terms, as a kind of fortuitous meeting place between a bodily hardware and a mental software susceptible of migration to "the cloud".

4.1 Metaverse: What Are We Talking About?

Having established the renewed status of the environment, we can say something more about the most paradigmatic—at the moment—example of the so-called 'metaverse'. As discussed, we often inhabit non-extensive places and we depend on them in many dimensions of our existence; from work to social life, from family to public relations, everything passes through the filter, for example, of the internet or digital platforms. The smartphone has become, for all of us, an extension of the arm and often also of our mind. Even more extreme situations occur with augmented reality and virtual reality, which insert a constant and continuous interaction between the dimension of reality and virtual parallel worlds.

In this sense, the metaverse seems to represent a *liberation from space* which has passed through the de-structuring of activities previously located within rigid perimeters based on the Aristotelian units of space-time-action. All this leads us to ask again with (Heidegger, 2001): what does space become today and what are the capabilities necessary to inhabit it and, therefore, to build it? In redefining the idea of space, technology and the metaverse play a crucial role as reality is aimed elsewhere, no longer located. But in any form, it substitutes what we know as 'environment'. The metaverse, in fact, while apparently healing the classic Cartesian body-mind/hardware-software dualism, as a simulation of some environment's features, cannot be considered an alternative to it. Metaverse's value relies upon the kind of relationships it is able to represent and leaves many open questions regarding its ability to make us experience the environment and human relationships. As it stands, it seems to be more extractive that enabling of human relationships, with other or with the real environment. The difference is not, however, in the available technology but in the virtues, habits, social practices and human expectations that push people to use the metaverse technology.

Consistently with what we have said above, without people an AI systems and virtual contexts immediately stop working, AI no longer 'understands' or even 'simulate' anything. No virtual reality, no *metaverse*—and therefore no *digital environments*—actually exists by themselves either. Again, a digital representation of reality exists—of the only reality that exists—within which the virtual has a place as a representation, but decays when there is no one to represent something to. In this sense -we agree with Luc Julia-, AI does not exist. It is the mere effect of a poor imagination, that can unfortunately be placed at the service of economic or political ambition.

The barriers that the metaverse is therefore helping to overcome is not space and time as such, but the strictly link we previously experienced of a one-to-one relationship between function and space. In doing so, the conventional closed and delimited space can evolve into an open, multifunctional space, intended for a fluid and plural use, which aims to define new places and new ways of inhabiting a post-topic world-environment while always maintaining the existential structure of human relationality. Functional goals a therefore still mediated by the AI technology, not question about meaning, opportunity, relevance, justice.

5 Sustainable AI

"Bad theories destroy good practices," (Ghoshal, 2005) said. The world of the digital innovation is no exception. We are seeing things that work (from digital solutions to the metaverse, from open spaces to phygital meetings), but the reason for their effectiveness and above all the conditions of sustainability of these innovation processes may not hold in virtue of such technological advancements. As said, supporting them without reflecting on the epistemological and axiological foundations means running into the danger denounced by Ghoshal and which is already a reality, for example, in the seven sins attributed to Greenwashing (CIT…): the sin of omission of information, the sin of no proof, the sin of vagueness, the sin of irrelevance, the sin of the lesser evil, the sin of lying, of adopting false labels. In this sense, a superficial technological innovation is possible, as a superficial ecologism is.

A sustainable AI technology development is possible on the basis of human choices and responsible technology's use. The pervasiveness of the digital tools asks for a deeper reflection about human relationships and desire. A different model of 'progress' is fuelled with narratives that do perpetuate wrong socio-imaginaries and toy stories as discusses. The idea that machines can at some point substitute human relationships. We can choose surrogate of them: this will be the problem, not that we can generate surrogates. We do so all the time, from food to clothes and even entertainment and art.

Narratives and toy stories affect human beings' expectations and prevent a serious process of personal and collective awareness and responsibility about what is at stake in the digital transitions. An innovation projected towards growth that reiterates productive and technological innovation like those that created the problems we suffer, systematically leads to avoiding reflection on limits and boundaries, which are instead generative of new practices and new economies. Even when we emphasize the importance of relationships and interdependencies, it seems easy to forget that taking them seriously means adapting, changing pace, dealing systematically with mutual dependencies where what actually make the difference are the human values, real relationships, human decisions and judgments.

In this sense, innovation is sustainable as far as it aims at the human flourishing, or it is not innovation at all. Its sustainability grounds in the ontology of the human beings, of the real world as our given environment and of AI as a tool. Such sustainability can be pursued as far as we take care of the motivations that generates, enhance AI technologies and encourage its use, not through a mere calculation of risks and of possible impacts. It is a social and educational problem, not a technical one. Again, it is the result of choices not of necessity.

Going back to the notions of environment and human ecology, the anthropological reason of what we are saying here is that any true innovation has its objective in the caring of other and of their place (environments). Places that should be understood as spatio-temporal experiences of co-existence with us ourselves, with others and with the world we inhabit. We have an obligation to care, to take care of ourselves, of others and of the ecological issues. It is a *human ecology* challenge.

For true innovation it is necessary to develop a new reasoning on human duties (even first than on human rights). Such duties exist in everyone's conscience even if no one were to recognize them, Simon Weil (Weil, 2017) would say. For sustainable innovation, we do not need regulations in the first place, but a new personal and collective moral and ethical conscience.

5.1 Prudential Intelligence

Moreover, when dealing with *an uncertain world*, we have to use not only the brute force of algorithms, but also the imagination, creativity, intuition and prudence human beings are capable of. In fact, the most comprehensive "method", the one that regulates the application of all the others, including the automatable ones, is human prudence.

As far as we know, the universe is not some kind of eternal clock, but a unique, historical, and contingent event. It is endowed with a certain network of regularities, sufficient to make life and intellection possible, but compensated with unpredictable novelties. This very peculiar distribution of constancy and rupture affects both the orbit of the planets, as well as our daily life, made of imperfect cycles, circadian rhythms, habit and shock. Only a living, sentient, located and interactive intelligence, a prudent intelligence like the human, can understand this disconcerting texture of the universe. We learn from experience, of course, but we know at the same time that there is no guarantee that things will continue as they were. Hence the convenience of intellectual humility, which has been dressed over time as a Socratic attitude, Aristotelian prudence, learned ignorance, fallibilism…

An AI system generates expectations. (It places a point in an n-dimensional space constructed from a history of data, and, based on it, it tells us what can be expected regarding the object represented by that point.) But the system can crash when it registers the occurrence of something whose possibility was not even considered in advance. When this happens, the system itself is left without the ability to adapt, it cannot learn from this experience. When this happens, it is not the algorithms that must react, but the people responsible for them. And they will react, first of all, by drastically changing expectations. They can do it since they are not artefacts, but conscious people who can come to understand the phenomenon which is not expected from the machine. Humans can activate their creativity to generate better expectations from now on with or without mechanical help. A person can conjecture causal relationships, beyond the mere correlations that a machine detects. And this step does not have to be purely arbitrary, random or irrational, but rather, in some sense, it is guided by a practical and social knowledge that Aristotle called *phrónesis*, prudence. Said knowledge facilitates (*i*) the integrative constitution of the experience, (*ii*) the management of emotions linked to the frustration of expectations, (*iii*) the propaedeutics of the creative moment and (*iv*) the critical filtering of the new emerging expectations.

In the words of Erik Larson: "AI works on inductive reasoning, crunching data sets to predict outcomes. But humans don't correlate data sets: we make conjectures informed by context and experience. Human intelligence is a web of best guesses, given what we know about the world. We haven't a clue how to program this kind of intuitive reasoning, known as abduction [(Aliseda, 2010)]. Yet it is the heart of common sense [prudence]. That's why Alexa [or SIRI, or ChatGPT] can't understand what you are asking".

Consequently, an AI system that aims to replace human prudence would simply be out of place, out of the universe that houses us (out of our uncertain world). On the contrary, a DeCo system inscribed within the framework of the prudential human intelligence will be in its rightful place and will be able to fulfil functions of a great value for human life.

6 Final Practical Orientations

The practical problems have, thus, nothing to do with a supposed post-human future of intelligent machines. "What should terrify us -says Ramón López de Mántaras, founder of the CSIC's AI Research Institute – is not a future dominated by a hypothetical superior AI […] What should really worry us is the present situation, in which we are delegating more and more tasks in an AI as limited as the current one" (López de Mántaras, 2020 [59]). And "delegating" is the key word here. What is crucial has to do with the present, with the way in which DeCo systems are already being used, with the responsibility that certain people, companies and governments have for it, as well as with the impact that this use already has on our life.

The new name (DeCo systems (Bertolaso & Marcos, 2023)) does justice to the true ontology of these systems and, above all, leads us more directly to important practical issues. We begin to see what the relative position of humans and machines should be. It is not a matter of deforming the former, the humans, to fit into a world presumably dominated by mechanical intelligences, but of placing the latter, the machines, within the framework of human life. Outside this framework, as we said, they cease to function, even cease to be what they were. What algorithms can contribute to human life? They allow us to delegate the control of certain processes.

Only at this point we can more appropriately raise the pertinent practical issues, ethical, political, educational, legal… Whose is the hand that rocks the algorithms? Who delegates? Are they entitled to do so? In which DeCo systems do they delegate? Are they the most appropriate systems? What kind of actions/processes are delegated? Are they really delegable?, technically?, ethically? For how long is control delegated? Is it sensible? Is the delegation reversible? What monitoring or evaluation procedures exist? Are they sufficient? What risks are assumed in case of failure? Is it prudent to assume them? What advantages are obtained for human life with the delegation of control? What do we lose in return? …

At the educational level, it is, of course, essential to teach young people (and not so young) certain values and train them in virtue, rather than insist on instructing machines in the ideology of political correctness. Unfair biases will not be redirected just by redesigning algorithms, but by educating virtuous people. However, virtues and values have to be educated to be at the level of the current technological context, as a part of a new *technological humanism*. For example, through intermittent technological silence practices (Marcos, 2020).

At the political level, the DeCo approach allows us to immediately identify the legitimacy deficit and the risk to people's freedom. Thus, most of the DeCo technologies are ultimately in the hands of the Chinese Communist Party and a few large US corporations. The greats of the so-called AI are, on the Western side, the GAFAM (Google, Amazon, Facebook, Apple and Microsoft) and, on the Chinese side, the BATX (Baidu, Alibaba, Tencent, Xiaomi, and you can add now Tik Tok or Huawei). The first impression we get is that there is too much power in too few hands.

We have to ask ourselves if the power they hold is legitimate. If it is not, neither its delegation to algorithms will be. It is imperative, for the health of democracies and for the freedom of the people, that the excessive power of these corporations be dissolved. Responsible digital consumption, lucid use of social networks and prudent management of each one's own data would already help (Véliz, 2020). But in addition, political pressure and public opinion in favour of the dissolution of nuclei of abusive digital power would be pertinent.

Just one example of something that is happening right now, in the last few days. Google has recently developed a tool, *Genesis*, designed specifically for the media, for the preparation of news for the media. They have already presented this application to the main North American media, *New York Times*, *Wall Street Journal*, and *Washington Post*. Through this software, Google could obtain a determining influence in the construction of world public opinion. We cannot leave the construction of the media vocabulary and the media agenda simply in the hands of Google. We don't want *Genesis* to become *Apocalypse*. At least, it is essential to demand a public clear labelling of the media pieces built by an AI system.

The new European AI law already takes steps in the right direction. The USA, UK and the G-7 have also taken steps, in recent days, towards the legal control of AI. Even the UN has recently issued a document with recommendations on the governance of AI.[1] However, as AI is now in the hands of half a dozen US corporations and the Chinese Communist Party, a determined activism in favour of the fragmentation of digital power is required. It must be a simultaneous task to the proclamation of laws. Without addressing it, any restrictive European legislation will only serve to make Europe even more irrelevant in the AI concert. Perhaps also

[1] See https://www.un.org/sites/un2.un.org/files/ai_advisory_body_interim_report.pdf (accessed May 12, 2024).

to relocate the most disturbing or dangerous applications of the AI to the most vulnerable countries.

Let's finally go back to the so-called digital rights. It should be clear by now that robots have no rights, nor are they responsible for anything, nor do they have to pay taxes, that it is absurd to talk about software rights. We are before a category mistake. But neither does it seem very perceptive or very useful to ask for a new generation of human rights, this time digital rights –let's say. Inventing new human rights weakens the very idea of "human rights". Human rights are based on the dignity of the person and their belonging to the human family. It is always tempting to use this formula –"human rights"—to protect any asset considered valuable. But, as the field to which this formula is applied expands, its protective force inexorably diminishes.

If we understand by human rights those of the first generation, fundamentally the right to life and liberty, then the accusation of having violated human rights is an extremely serious accusation. But if we include not only second and third generation rights, but also new digital rights and neurorights, plus the "human" rights of robots or animals, then the accusation of violation of a right becomes slightly disturbing. Much more insightful and useful would be to connect the digital to the basic human rights, to show how it affects them, without inventing new lists of supposed human rights. What is serious about some DeCo systems is not that they violate our alleged digital rights, it is that they can sometimes threaten our lives or compromise our freedom.

Acknowledgments We thank Laura Corti and Maria Laura Ilardo who are contributing to organizing some reflections of the first paragraphs of the current paper.

References

Aliseda, A. (2010). *Abductive reasoning*. Springer.

Bertolaso, M., & Marcos, A. (2023). *Umanesimo tecnologico*. Carocci.

Capone, L., Rocchi, M., & Bertolaso, M. (2023). *Rethinking "digital": A genealogical enquiry into the meaning of digital and its impact on individuals and society*. AI & Society.

Ghoshal, S. (2005). Bad management theories are destroying good management practices. *Academy of Management Learning & Education, 4*(1), 75–91.

Heidegger, M. (2001). *Fundamental concepts of metaphysics: World, finitude, solitude*. Indiana University Press. (new edition).

Heidegger, M. (2008). *Being and time*. Harper Collins. (reprint edition).

Hirschheim, R., Klein, H. K., & Lyytinen, K. (1995). *Information systems development and data modelling*. CUP.

Johnson, B. (1967, October 9). *Speech at Yale University, white house diary*. Retrieved September 8, 2024, from https://www.nps.gov/common/uploads/teachers/lessonplans/Lady%20Bird%20Redefines%20Beauty%20Curriculum1.pdf

Julia, L. (2019). *L'intelligence artificielle n'existe pas*. Éditions First.

Larson, E. J. (2021). *The myth of artificial intelligence. Why computers can't think the way we do*. Harvard University Press.

Leopold, A. (2020). *A Sand County almanac: And sketches here and there*. Oxford University Press. (illustrated edition).

López de Mántaras, R. (2020, July). El traje nuevo de la IA. *Investigación y Ciencia, 526*, 50–59.
Marchesini, R. (2002). *Post-human: verso nuovi modelli di esistenza*. Bollati Boringhieri.
Marcos, A. (2020). New praxis. Releasement in a technological world. In L. Valera & J. C. Castilla (Eds.), *Global changes. Ethics, politics and environment in the contemporary technological world* (pp. 17–27). Springer.
Naess, A. (2016). *Ecology of wisdom*. Penguin Classics.
Peirce, C. S. (1935). *Collected papers* (Vol. 5). Harvard University Press.
Shannon, C., & Weaver, W. (1963). *The mathematical theory of communication*. University of Illinois Press.
Uexkull, J. (2010). *A foray into the worlds of animals and humans: With a theory of meaning*. University of Minnesota Press. (illustrated edition).
United Nations (UN). (1987). *Report of the world commission on environment and development: Our common future. transmitted to the general assembly as an annex to document A/42/427–development and international co-operation: Environment*. Retrieved May 12, 2024, from https://www.unicas.it/media/2732719/Rapporto_Brundtland_1987.pdf
Véliz, C. (2020). *Privacy is power*. Transworld.
Von Hopffgarten, A. (2021). La inteligencia artificial carece de la flexibilidad de decisión humana. Interview with Katharina Zweig. *Mente y Cerebro*, enero/febrero.
Weil, S. (2017). *La prima radice* (F. Fortini, Trans.). Edizioni di Comunità.

Index

A. Fabris, S. Belardinelli (eds.), *Digital Environments and Human Relations*, Human Perspectives in Health Sciences and Technology 150,
https://doi.org/10.1007/978-3-031-76961-0

O

P

R

S

U

V

GPSR Compliance

The European Union's (EU) General Product Safety Regulation (GPSR) is a set of rules that requires consumer products to be safe and our obligations to ensure this.

If you have any concerns about our products, you can contact us on ProductSafety@springernature.com

In case Publisher is established outside the EU, the EU authorized representative is:

Springer Nature Customer Service Center GmbH
Europaplatz 3
69115 Heidelberg, Germany

The manufacturer's authorised representative in the EU is Springer Nature Customer Service Centre GmbH, Europaplatz 3, 69115 Heidelberg, Germany. If you have any concerns regarding our products, please contact ProductSafety@springernature.com

Printed and bound by CPI Group (UK) Ltd, Croydon, CR0 4YY
30/09/2026
02225564-0001